Complex Carbohydrates

Their Chemistry, Biosynthesis, and Functions

Complex Carbohydrates

Their Chemistry, Biosynthesis, and Functions
A Set of Lecture Notes

Nathan Sharon

The Weizmann Institute of Science

1975
Addison-Wesley Publishing Company
Advanced Book Program
Reading, Massachusetts

London · Amsterdam · Don Mills, Ontario · Sydney · Tokyo

574.19248
Sh 2
c.

First printing, 1975
Second printing, 1978

Library of Congress Cataloging in Publication Data

Sharon, Nathan.
  Complex carbohydrates, their chemistry, biosynthesis, and functions.

  An updated and expanded version of lecture notes prepared for students at the Feinberg Graduate School of the Weizmann Institute of Science, Rehovoth, Israel.
  1. Carbohydrates. I. Title.
QP701.S477   574.1'9248   75-31746
ISBN 0-201-07324-2
ISBN 0-201-07323-4 pbk.

Reproduced by Addison-Wesley Publishing Company, Inc., Advanced Book Program, Reading, Massachusetts, from camera-ready copy prepared by the author.

Copyright © 1975 by Addison-Wesley Publishing Company, Inc.
Published simultaneously in Canada.

All rights reserved. No part of this publication may be reproduced, stored in a retrieval system, or transmitted, in any form or by any means, electronic, mechanical, photocopying, recording, or otherwise, without the prior written permission of the publisher, Addison-Wesley Publishing Company, Inc., Advanced Book Program, Reading, Massachusetts 01867, U.S.A.

Manufactured in the United States of America
ISBN 0-201-07323-4
CDEFGHIJKL-MA-8987654321

To Rachel with love

CONTENTS

	PREFACE	xiii
	ABBREVIATIONS	xvii
1	INTRODUCTION	1
	Materials for energy and structure	3
	Chemical problems	6
	Ups and downs in interest	9
	New and unusual monosaccharides	13
	Sugars linked to nucleotides and lipids	18
	New control mechanisms	23
	Genetic diseases	24
	Carbohydrates as determinants of biological specificity	26
2	GLYCOPROTEINS - I: GENERAL	33
	Sugar constituents	42
3	GLYCOPROTEINS - II: ISOLATION AND CHARACTERIZATION	48
	Problems of sugar analysis	52
	Mechanism of hydrolysis of glycosides	56
	The hydrolysis by acid of glucosaminides	59

4	CARBOHYDRATE-PEPTIDE LINKAGES	65
	The N-acetylglucosaminyl-asparagine linking group	69
	Protection of sugars against alkali by borohydride	72
	O-Glycosidic linkages to serine and threonine	73
	The galactosyl-hydroxylysine linkage	78
	The arabinosyl-hydroxyproline linkage	79
	Unusual linkages	80
5	GLYCOPEPTIDES - I: ISOLATION AND CHARACTERIZATION	84
	Use of glycosidases	87
	Chemical techniques: methylation and Smith degradation	91
6	GLYCOPEPTIDES - II: STRUCTURAL FEATURES	99
	Microheterogeneity	99
	Common structural features	103
	The Asn-X-Ser(Thr) sequence	108
	Mucins	111
	Antifreeze glycoproteins	112
	Collagen and basement membrane	113
7	GLYCOPROTEIN BIOSYNTHESIS - I	118
	Studies with intact cells	119
	Biosynthesis of immunoglobulins	127
	Synthesis of sugar constituents	128
	Biosynthesis of 6-deoxy sugars	131
	Biosynthesis of galactose and xylose	138
8	THE SIALIC ACIDS	142

CONTENTS

9	GLYCOPROTEIN BIOSYNTHESIS - II	155
	Biosynthesis of saccharide side chains in cell free systems	155
	Specificity and distribution	159
	Control of glycoprotein biosynthesis	161
	Specification of anomery	166
	Feedback control mechanisms in glycoprotein biosynthesis	168
	Role of lipid linked intermediates	169
10	FUNCTIONS OF THE CARBOHYDRATE - I	177
	Postulated functions of protein-bound carbohydrates	178
	Physico-chemical properties of the glycoprotein	178
	Sugars and the interaction between macromolecules and membranes	182
	Carbohydrates in membrane-membrane interactions	188
11	FUNCTIONS OF THE CARBOHYDRATE - II	193
	Sugars and the survival of glycoproteins in circulation	193
	Liver membrane receptors	204
	Cell surface saccharides and lymphocyte activation	207
12	BLOOD GROUP SUBSTANCES - I: CHEMICAL STRUCTURE	215
	The ABH and Lewis antigens	218
13	BLOOD GROUP SUBSTANCES - II: GENETIC CONTROL AND BIOSYNTHESIS	235
	Glycosyltransferases specified by the blood group genes	239
	The H gene enzyme	240
	The Le gene enzyme	241
	The A gene enzyme	242
	The B gene enzyme	245

	Blood group M and N specificities and the major glycoprotein of the human erythrocyte membrane	248
14	MUCOPOLYSACCHARIDES (PROTEOGLYCANS) - I: CHEMICAL STRUCTURE	258
	Hyaluronic acid	266
	Chondroitin sulfates	268
	The linkage region	270
	Keratan sulfate	272
	Structural heterogeneity of repeating disaccharide units	273
	Structure of native proteoglycans	276
	Enzymic degradation of mucopolysaccharides	278
15	MUCOPOLYSACCHARIDES (PROTEOGLYCANS) - II: BIOSYNTHESIS	282
	Biosynthesis of chondroitin sulfate	283
	Xylosyl transfer	284
	Galactosyl transfer	287
	Glucuronyl transfer to galactose	289
	Repeating disaccharide formation	289
	Participation of multienzyme complexes	291
	Sulfation	292
	Biosynthesis of $\underline{L}$-iduronic acid	295
	Lipids in mucopolysaccharide biosynthesis	298
	Regulation of mucopolysaccharide biosynthesis	299
16	GENETIC DEFECTS OF MUCOPOLYSACCHARIDE METABOLISM	303
	Cell culture studies	307
	Identification of the missing enzymes	310
	Practical benefits	314

CONTENTS xi

17	THE BACTERIAL CELL WALL	318
	Gram-positive bacteria	322
	Gram-negative bacteria	328
	The biologically versatile lipopolysaccharides	331
	Composition and structure	336
18	LIPOPOLYSACCHARIDES - I: STRUCTURE AND BIOSYNTHESIS	345
	Use of defective mutants	345
	Structure of the O-specific chains	352
	Lipid A	355
	Biosynthesis of the core polysaccharide	356
	Role of phospholipids in sugar transfer	363
19	LIPOPOLYSACCHARIDES - II: BIOSYNTHESIS	369
	Biosynthesis of O-specific chains	369
	Mechanism of polymerization	375
	Direction of chain growth	378
	Joining the parts	381
	Assembly of the outer membrane	382
	Phage conversion of O-antigen	384
	Metabolic control of LPS biosynthesis	390
20	PEPTIDOGLYCAN - I: STRUCTURE	393
	Structure of the glycan moiety	396
	Structure of the peptide moiety	404
	Peptidoglycan chemotypes in different bacterial species	411
	General structure of the peptidoglycan	415

21	PEPTIDOGLYCAN - II: BIOSYNTHESIS AND MODE OF PENICILLIN ACTION	418
	Synthesis of precursors	420
	Assembly of the disaccharide-peptide repeating unit	426
	Modification of the peptide side chain; role of tRNA	430
	Polymerization of disaccharide units	435
	Crosslinking of peptide units	436
	Role of the transpeptidase in peptidoglycan biosynthesis	439
	Transpeptidase and DD-carboxypeptidase	443
	Architectural relationships in peptidoglycan synthesis	448
INDEX		453

PREFACE

This book presents an updated and expanded version of lecture notes that I started to prepare early in 1974, in the aftermath of the October 1973 Middle East War. The original purpose of the notes was to serve as reading material for the students of the Feinberg Graduate School of the Weizmann Institute, who at the time were still mobilized in the Israel Defence Forces and could not attend my course on the Biochemistry of Complex Carbohydrates. Keeping in mind the conditions under which the students were to consult the notes, I attempted to make the material simple, clear and highly comprehensible, but not necessarily comprehensive, and to go beyond established facts to present newer concepts. What is perhaps more important, I tried to convey the feeling of excitement and fascination which, I believe, permeates research on carbohydrates, which is now undoubtedly one of the liveliest and most exciting fields in biochemistry.

The need for such notes was also prompted by the fact that no suitable textbook or monograph on the subject has been available, most of the material covered in my course being scattered in reviews and original articles. References to these are given at the end of each lecture.

In their present form, the notes are aimed mainly at graduate students in biochemistry and related areas. I do hope, however, that they will also be of aid to both young and established researchers in carbohydrate biochemistry. Only a basic knowledge of the chemistry of carbohydrates is assumed. Because of shortage of time and space, the coverage is not complete. I have not dealt with broad topics such as glycolipids, nor more specific ones such as lipoteichoic acids and yeast mannans.

In the preparation of the lecture notes for publication, I have been greatly encouraged by the many students, friends and colleagues, who not only shared with me the feeling that there is a pressing need for such a book, but also helped me with their comments and suggestions. I am particularly indebted to my colleague Dr. Halina Lis, who gave much of her time and thought to help me in the prepar-

PREFACE   xv

ation of the final manuscript. I wish also to express my gratitude to my highly efficient and dedicated secretary, Mrs. Dvorah Ochert, who painstakingly typed the manuscript several times and introduced many useful corrections. Without their efforts, I doubt if this book could have been completed.

N.S.

ABBREVIATIONS

Ac	acetyl
ACL	antigen carrier lipid
ADP	adenosine diphosphate
AMP	adenosine monophosphate
ATP	adenosine triphosphate
BSM	bovine submaxillary mucin (glycoprotein)
CDP	cytidine diphosphate
CMP	cytidine monophosphate
CTP	cytidine triphosphate
DAP	diaminopimelic acid
DNA	deoxyribonucleic acid
Ea	ethanolamine
EC	extracellular space
EDTA	ethylene diamine tetraacetate
FSH	follicle stimulating hormone
GalUA	galacturonic acid
GCL	glycosyl carrier lipid
GDP	guanosine diphosphate
GDP-4KDM	GDP-4-keto-6-deoxymannose

Glc	glucose
GlcN	glucosamine
GlcNAc-Asn	*N*-acetylglucosaminyl-asparagine
GlcUA	glucuronic acid
GTP	guanosine triphosphate
HCG	human chorionic gonadotropin
Hep	heptose (L-glycero-D-mannoheptose)
IdUA	L-iduronic acid
KDO	2-keto-3-deoxyoctanoic acid
LH	luteinizing hormone
LPS	lipopolysaccharide(s)
MGT	multiglycosyltransferase(s)
MurNAc	*N*-acetylmuramic acid
NAD	nicotinamide adenine dinucleotide
NADH	reduced NAD
NADP	nicotinamide adenine dinucleotide phosphate
NADPH	reduced NADP
NANA	*N*-acetylneuraminic acid
NGNA	*N*-glycolylneuraminic acid
OSM	ovine (sheep) submaxillary mucin (glycoprotein)
PAP	3'-phosphoadenosine-5'-phosphate
PAPS	3'-phosphoadenosyl-5'-phosphosulfate
PEa	phosphatidylethanolamine
PEP	phosphoenolpyruvate
P-GCL	phosphate-linked glycosyl carrier lipid (bactoprenyl phosphate)
PHA	phytohemagglutinin from *Phaseolus vulgaris*
P_i	inorganic phosphate
PM	plasma membrane
PP-GCL	pyrophosphate-linked glycosyl carrier lipid (bactoprenyl pyrophosphate)

ABBREVIATIONS

PP_i	inorganic pyrophosphate
PSM	porcine submaxillary mucin (glycoprotein)
RER	rough endoplasmic reticulum
RNA	ribonucleic acid
SDS	sodium dodecyl sulfate
SG	secretory granules
SR	semirough
TDP	thymidine diphosphate
TDP-4KDG	TDP-4-keto-6-deoxyglucose
tRNA	transfer ribonucleic acid
TTP	deoxythymidine triphosphate
UDP	uridine diphosphate
UMP	uridine monophosphate
UTP	uridine triphosphate
WGA	wheat germ agglutinin

Complex Carbohydrates

Their Chemistry, Biosynthesis, and Functions

INTRODUCTION

To an observer trying to obtain a bird's eye view of the present state of biochemistry, or what is sometimes referred to as molecular biology, life may until very recently have seemed to depend on only two classes of compounds: nucleic acids and proteins. We know, however, that living organisms cannot exist without other classes of compounds, notably carbohydrates and lipids. One of these, the carbohydrates, is the subject of my lectures.

Carbohydrates form a very large group of compounds of enormous theoretical and practical importance, both from the quantitative and qualitative points of view. Research on these compounds has undergone an immense revitalization during the last decade, and is continuing to expand very rapidly. As a result of many new developments this field is now broad in scope and rich in variety, touching on virtually all aspects of chemistry and biology.

The study of carbohydrates and their derivatives has enriched our knowledge of chemistry, especially as to the role of molecular shape and conformations in chemical reactions. It has led to the discovery of new biosynthetic re-reactions and enzymic control mechanisms and has contributed significantly to the understanding of many fundamental biological processes, such as the interactions of cells with their environment. Moreover, it has provided a basis for the recognition of the enzymic defects of several genetic disorders, and has raised hopes that it may be possible to treat them effectively.

We shall not attempt to survey the field comprehensively, but rather discuss several major aspects of it with emphasis on recent advances related to the structure, biosynthesis and function of complex saccharides. Much of the material to be covered is not found in textbooks, even the most recent ones. I have therefore included references to reviews and original articles which will help you in your studies. Basic carbohydrate chemistry will not be included, as it is well covered in standard textbooks (1-3); nor shall I deal with conformational analysis of sugars (4), a subject of ut-

INTRODUCTION

most importance for the understanding of the chemical as well as the biological reactions of sugars (5). I urge you to consult some of the books and articles listed in the references. As an introduction to the course, I strongly recommend that you read the chapters on carbohydrates in any advanced textbook of biochemistry.

The name carbohydrate was originally assigned to compounds believed to be hydrates of carbon, of the general formula $C_n(H_2O)_n$. With the accumulation of knowledge the definition has been modified and broadened. It now includes polyhydroxy aldehydes and ketones, alcohols and acids, their simple derivatives as well as the products formed by the condensation of these different compounds via glycosidic (hemiacetal) linkages into oligomers (oligosaccharides) or polymers (polysaccharides). In fact many compounds of unusual structure, which do not conform to the general formula $C_n(H_2O)_n$ are now included in the enlarged group of carbohydrates.

MATERIALS FOR ENERGY AND STRUCTURE

The amount of carbohydrates found in nature is larger than that of any other group of natural compounds. The most abundant organic substance on earth is cellulose, a polymer

of glucose*, which is the structural material of plants.
Another very abundant substance is chitin, a polymer of N-acetylglucosamine. Chitin is the major organic component of the exoskeleton of arthropods such as insects, crabs and lobsters. Since arthropods are the largest class of organisms, comprising some 900,000 species (more than are found in all other families and classes together), it is little wonder that chitin is found in nature in very large quantities. It has been estimated that millions of tons of chitin are formed yearly by a single species of crab.

Carbohydrates are also the main source of energy for living organisms and the central pathway of energy supply in most cells. They are the major products through which the energy of the sun is harnessed and converted into a form that can be utilized by man and other animals, as well as by many other organisms. According to rough estimates, some 4×10^{11} tons of carbohydrates are being formed each year on earth by the process of photosynthesis. The starches and glycogens, long chain polymers of glucose the structure of which differs from that of cellulose, are the media for energy storage in plants and animals, respectively. Coal, peat and petroleum

---

*All sugars mentioned are of the $\underline{D}$-configuration, unless otherwise noted. For simplicity the symbol $\underline{D}$ is usually omitted.

INTRODUCTION

were most probably formed from carbohydrates by microbiological and chemical processes.

Carbohydrates are important constituents of the human diet and comprise a high percentage of the calories consumed. Thus, some 50-60% of the caloric intake of the inhabitants of Israel is in the form of carbohydrates - glucose, fructose, lactose, sucrose and starch. Sucrose is the major sugar of our food; its world production is now over 60 million tons a year. Carbohydrates also serve as the basis for industries of great economic importance such as paper and pulp, textile fibers and pharmaceuticals. The principal industrial carbohydrate is undoubtedly cellulose; its world-wide usage is estimated at 800 million tons per year.

The polysaccharides just mentioned - cellulose, glycogen, starch and chitin - are relatively simple polymers: they are homopolymers, made up of one type of monomer, glucose or N-acetylglucosamine. This seeming simplicity, perhaps even dullness of structure, is probably one of the reasons why biochemists for a number of years lost interest in carbohydrates. You can see a reflection of this lack of interest by looking through any textbook of biochemistry, where carbohydrates are relegated to a secondary place, after proteins and nucleic acids, and where they are discussed solely as structural or protective materials and as energy sources. Even in the most modern textbooks there is very little mention, if any, of the many other essential roles that carbohydrates perform, some of which I shall discuss in detail in this course.

CHEMICAL PROBLEMS

Another important reason why biochemists shied away from the study of carbohydrates stemmed from the many chemical problems encountered in dealing with these materials. Sugars are multifunctional compounds, with a number of hydroxyls, most commonly four or three, of approximately equal chemical reactivity. Manipulation of a single selected hydroxyl group is in many cases a serious problem to this very day. Blocking one hydroxyl group, or leaving one free, can be achieved only with great difficulty, and requires careful design and execution of a complex series of reactions. Synthesis of a disaccharide is therefore a considerable achievement, trisaccharides have rarely been synthesized, and there are only very few reports on the synthesis of higher oligosaccharides. This is in marked contrast to the situation in peptide chemistry, where peptides made up of dozens of amino acids can readily be synthesized, not only manually

INTRODUCTION

but also by automatic techniques, and at least three proteins - insulin, made up of 51 amino acids, ribonuclease (124 amino acids), and lysozyme (129 amino acids) - have already been synthesized. One reason for the relative ease of such syntheses is that the number of steps involved in the preparation of a peptide is rather small and is considerably less than that required for the synthesis of an oligosaccharide of similar size. Even more important is the tremendous difference in the number of isomeric oligopeptides and oligosaccharides that can be obtained from the same number of corresponding monomers (Table 1).

Table 1

Comparison of the number of possible isomeric peptides and oligosaccharides^a

Composition	Product	Number of isomers	
		Peptides	Saccharides^b
X_2	dimer	1	11
X_3	trimer	1	176
XYZ	trimer	6	1056

^{a}Calculated by John Clamp, University of Bristol
^{b}Pyranose ring only

From one amino acid, such as glycine, alanine or proline, we can make only one dipeptide. However, from one hexose, such as glucose in its pyranose (six membered ring) form, we can make eleven different disaccharides. Eight of these are obtained by forming a glycosidic linkage via the oxygen atom attached to C-1, the anomeric carbon, of one sugar ring either in α or β configuration to carbons 2,3,4 or 6 of another sugar ring. Carbon 5 is not free to participate in the formation of the glycosidic linkage since it is involved in formation of the pyranose ring. The remaining three disaccharides are formed by linking C-1 of both sugar residues through a glycosidic oxygen in either αα, αβ or ββ linkages. The number is much greater if we also include disaccharides made from the furanose (five membered ring) form of glucose.

$\underline{D}$-Glucose is the only sugar for which the complete series of eleven pyranose disaccharides has been prepared. It is interesting to note, however, that only one of these isomers, α,α-trehalose (or α-glucosyl-α-glucoside) occurs, to any significant extent, free in nature. Trehalose is the major component of the blood sugar of adult insects, and is a reserve carbohydrate in these, as well as in many other organisms. Another disaccharide, gentiobiose (Glc-β-(1→6)-Glc), occurs as a simple glycoside. Most of the other nine disaccharides occur in polysaccharides from which they can

INTRODUCTION

be prepared by degradation, e.g. cellobiose (Glc-β-(1→4)-Glc) from cellulose; maltose (Glc-α-(1→4)-Glc) from starch.

If we now consider the trimers, a single amino acid will give us only one tripeptide, whereas from a single hexopyranose we can prepare - theoretically at least - 176 different trisaccharides. Finally, three different amino acids can form 6 tripeptides, whereas from the same number of hexoses the number of possible trisaccharides is 1056. The sugars in these saccharides, as in all other oligo- and polysaccharides known to us, are bound by glycosidic linkages.

An added complication which is not encountered in proteins and nucleic acids stems from the fact that whereas the latter are linear polymers, where branching is very rarely encountered, in polysaccharides branching is of common occurrence. This increases greatly the number of possible structures, and thus the difficulties in studying these compounds.

UPS AND DOWNS IN INTEREST

I have given you some of the reasons why carbohydrates were neglected by most chemists and biochemists for a long time, in fact until the early 1960's. Explicit expressions of this lack of interest have been recorded in the scientific literature. Thus, the British carbohydrate chemist D. J. Bell relates that when, in the 1930's, he embarked upon the study of carbohydrates, he was told by his seniors working on other subjects that he would be ill advised to do so "as the field had now been fully worked out". Only some twenty monosaccharides were then known; their number now is well over 100.

In 1971 D. A. Rees, in the Eighth Colworth Medal Lecture entitled "Shapely Polysaccharides" (6), had this to say about the situation in carbohydrate research a dozen years ago:

> " I moved to Edinburgh where carbohydrate research was, of course, very active already ... However, despite all the energy and thought that was going into our work, we all felt that carbohydrate chemistry and biochemistry were running down. Whelan mentions in his CIBA Lecture (in 1971) that he felt this way about starch and glycogen in about 1957. "

But then Rees goes on to say:

> " Times have changed, however, and great things have happened since then. We have seen a slow unfolding of the story of polysaccharide biosynthesis following Leloir's discovery of sugar nucleotides, which led to his Nobel Prize (in 1970). Likewise, with great pleasure, we have watched Strominger's progress with the relation between antibiotic action and the bacterial cell wall. The three-dimensional structure and probable mode of action of a carbohydrase enzyme has been shown to us by David Phillips and his group. Don Northcote and others have unravelled a great deal about how cytoplasmic organization leads to the structure of plant cell walls. The polysaccharides of mammalian connective tissue, and glycoproteins, begin to make biochemical sense for the first time ever. So many exciting dev-

INTRODUCTION 11

> elopments have occurred that this period seems
> to have moved us out of a dark age to see poly-
> saccharides in quite a new light. They have
> become interesting molecules to contemplate in
> relation to the life of the cell. The ugly
> ducklings have begun to look a little more like
> swans. In this sense, polysaccharides begin to
> appear attractive molecules, shapely molecules. "

The above is true nowadays not only for polysaccharides proper, but for complex saccharides in general, be they glycoproteins, glycopeptides, peptidoglycans or glycolipids.

If I may strike a philosophical note, this decline and rise of a scientific field is not unique to carbohydrates. After all, research in carbohydrates had its good days when it fascinated many great minds. The turn of the century witnessed Emil Fischer's trail-blazing carbohydrate work; some fifty years ago C. S. Hudson established his well known rules relating the optical properties of sugars to their structure and configuration, while W. N. Haworth devised the familiar sugar ring formula; thirty years ago M. L. Wolfrom was in the midst of his illustrious research on sugar chemistry and on the structural elucidation of polysaccharides, notably of heparin.

Even about nucleic acids it was once said:

> " ... one can only hope that if we have not
> reached the last chapter in an interesting ser-
> ies of researches which was initiated by Miescher
> many years ago, we must be somewhat near the
> penultimate one. "

It is hard to believe that this statement appeared in a review on the progress of chemistry written in 1909!

As mentioned, our attitude to carbohydrates has changed markedly during the last decade. There were many reasons for this change, first and foremost of which was undoubtedly the introduction of new and greatly improved techniques. Carbohydrate chemists working in the first half of this century had to rely almost exclusively upon carefully controlled chemical transformations and upon polarimetry in the investigation of the structures of monosaccharides and their derivatives. Work at that time was further restricted by the lack of good separation techniques. The advent of chromatography in its various forms, and of powerful instrumental analytical methods such as nuclear magnetic resonance spectroscopy, infrared spectroscopy, ultraviolet spectroscopy, mass spectrometry and X-ray diffraction analysis, has permitted a complete transformation in the approach to the problem of carbohydrate structure. Moreover, combinations of these techniques can provide information more rapidly, more conveniently, in greater detail and with smaller quantities of material than was previously possible.

INTRODUCTION

NEW AND UNUSUAL MONOSACCHARIDES

One result of the introduction of new and powerful techniques into carbohydrate chemistry was the discovery of many new saccharides, both simple and complex. In recent years there has been a dramatic increase in the number of rare sugars isolated from natural sources and these have provided the carbohydrate chemist with new and challenging problems of structural determination and synthesis. I shall give you one example from a field which I have been active in - new amino sugars (7). Until 1946 only two such sugars were known to occur in nature: glucosamine, discovered by Ledderhose in 1876, and galactosamine, discovered by Levene and LaForge in 1914. At the time it was believed that this was the end of the story, but this proved to be untrue as the following figures show:

Year	Number of known natural amino sugars
1946	2
1953	4
1960	20
1970	50

The first of the "new" amino sugars, discovered in 1946, was N-methyl-$\underline{L}$-glucosamine, a constituent of the antibiotic streptomycin. Subsequently many other new amino sugars were found in antibiotic substances (8). Indeed, some antibiotics have an oligosaccharide-like structure. These include the streptomycins and neomycins (Fig.1) as well as other aminoglycoside antibiotics such as the kanamycins and paromomycins, all of which are used clinically.

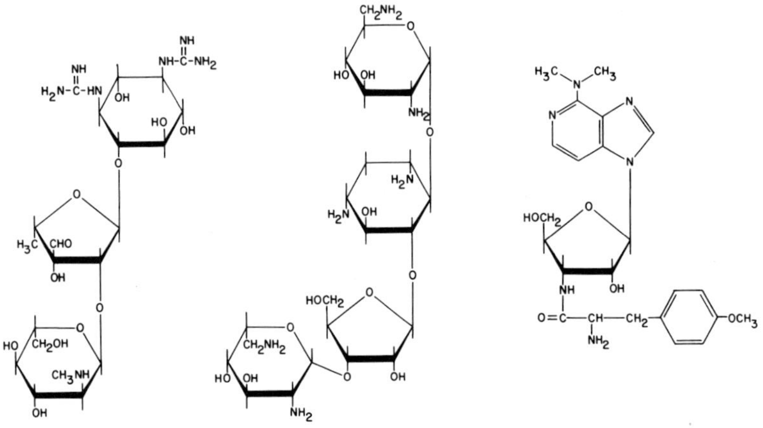

Streptomycin *Neomycin B* *Puromycin*

Figure 1 Aminoglycoside antibiotics. In streptomycin, the sugar ring at the bottom is of *N*-methyl-L-glucosamine. Neomycin B contains 2,6-diamino-2,6-dideoxy-L-idose (bottom ring) and 2,6-diamino-2,6-dideoxyglucose (top ring). 3-Amino-3-deoxyribose is the central ring in puromycin.

Neomycin B contains the diamino sugars 2,6-diamino-2,6-dideoxyglucose and its 5 epimer, 2,6-diamino-2,6-dideoxy-L-idose; kanamycin contains 3-amino-3-deoxyglucose

INTRODUCTION   15

and 6-amino-6-deoxyglucose. Another aminoglycoside antibiotic is puromycin, a well-known inhibitor of protein synthesis; it is a nucleotide analog in which 3-amino-3-deoxyribose is present instead of ribose. The antibiotics celesticetin and lincomycin (the latter therapeutically useful) contain as their carbohydrate moiety the unusual 8-carbon amino sugar, 6-amino-6,8-dideoxy-D-erythro-D-galactooctose (lincosamine) (8).

Lincosamine

$$\begin{array}{c} CH_3 \\ | \\ HOCH \\ | \\ H_2NCH \end{array}$$

To learn more about the mode of action of these antibiotics and to improve on them it is desirable to synthesize analogs with different amino sugar constituents. This has, of course, served as a strong impetus to the development of new methods of synthetic amino sugar chemistry.

New amino sugars, as well as other types of sugars, have been isolated in recent years not only from antibiotics but also from other sources, in particular from bacterial polysaccharides. One of the most important of these is the 3-O-D-lactic acid ether of glucosamine, known as muramic acid. This amino sugar, found only in bacteria, was isolated for the first time by R. E. Strange and F. A. Dark in England in 1956. Its N-acetylated derivative, N-acetylmuramic acid, together with N-acetylglucosamine form the polysaccharide backbone of the bacterial cell wall peptidoglycan.

N-Acetylmuramic acid

Another new sugar is ribitol (Fig.2), a reduction product of ribose. It is a constituent of the teichoic acids, discovered by J. Baddiley in England during the 1950's. The teichoic acids are polymers of ribitol-5-phosphate or glycerol phosphate, found in Gram-positive bacteria. In the cell walls of these organisms, they act as immunological determinants and as bacteriophage receptors.

An important sugar of unusual structure is neuraminic acid (Fig.2), the parent compound of the sialic acids, which

INTRODUCTION

```
CH₂OH              COOH              CHO               CH₂OH
 |                  |                 |                  |
HC-OH              C=O              HCNH₂              C=O
 |                  |                 |                  |
HC-OH              CH₂              HOCH               HOCH
 |                  |                 |                  |
HC-OH              HCOH             HCNH₂              HCOH
 |                  |                 |                  |
CH₂OH             H₂NCH             HCOH               HCOH
                    |                 |                  |
                  HOCH               CH₃              HOCH
                    |                                    |
                  HCOH                                 HCOH
                    |                                    |
                  HCOH                                 HCOH
                    |                                    |
                  CH₂OH                                CH₂OH

ribitol         neuraminic       bacillosamine      nonulose (from
                  acid                                avocado)
```

Figure 2 Some unusual sugars found in nature.

are widely distributed in nature. It occurs in glycosidically-linked form, mostly in glycoproteins and glycolipids, and in this form is a prominent constituent of cell surfaces. Neuraminic acid is a nine-carbon sugar acid, with an amino group in the molecule. It never occurs in nature unsubstituted: usually it is found in its N-acetylated form (N-acetylneuraminic acid, in brief NANA or NeuN), sometimes as the N-glycolyl (CH_2OH-CO-) derivative (NGNA), or as various disubstituted derivatives (e.g. N,O-diacetyl), all of which fall under the name "sialic acids". Sialic acid is also the constituent of the homopolysaccharide colominic acid found in *Escherichia coli*.

A rare diamino sugar - the first of its kind - which we have been studying in our laboratory for the last fifteen years, is bacillosamine. I discovered this sugar from a polysaccharide of *Bacillus licheniformis* (previously known as *Bacillus subtilis*) (A1) when I worked in the laboratory of R. W. Jeanloz at the Massachusetts General Hospital in Boston. Only recently, by the joint efforts of a number of coworkers, were we able to establish its structure, both by degradation and by chemical synthesis (A2) as 2,4-diamino-2,4,6-trideoxyglucose (Fig.2).

Finally, as a curiosity I would like to mention a rare nine carbon sugar, a nonulose ($\underline{\underline{D}}$-erythro-$\underline{\underline{L}}$-galactononulose, Fig.2), isolated from the avocado pear by Sephton and Richtmeyer in 1966 (A3). About 100 mg of the pure syrupy compound was obtained from 400 avocados, the total weight of which was 93 kg!

SUGARS LINKED TO NUCLEOTIDES AND LIPIDS (9)

I would now like to mention some other reasons for the increase of interest in carbohydrates. One of these, certainly of utmost significance, was the discovery of sugar nucleotides and their manifold roles as intermediates in biosynthetic reactions - whether of monosaccharides or of complex saccharides. The first sugar nucleotide, uridine diphosphate glucose (UDP-Glc), was discovered by Leloir and his coworkers in Buenos Aires in 1949.

Uridine diphosphate glucose (UDP-Glc)

To date, over 100 different sugar nucleotides have been identified. Most of these are of the general structure

XDP - Sugar

where X can be any of the five nucleosides - adenosine, guanosine, cytidine, uridine and deoxythymidine, and the sugar is one of a large variety of structures, some of which are very uncommon. One example of this type is cytidine diphosphate vinelose, isolated from *Azotobacter vinelandii*, the sugar moiety of which is branched (A4). Branched chain sugars constitute a rare group of natural compounds, some of which have been found in antibiotics and in plant glycosides.

Cytidine diphosphate vinelose (CDP-vinelose)

UDP-Glc is the sugar nucleotide found in highest concentrations in biological materials. It is the donor of glucose for the synthesis of glucosides (e.g. phenyl-β-glucoside), oligosaccharides (e.g. sucrose, lactose and trehalose), polysaccharides (e.g. starch and glycogen) and other glucose containing compounds. An unusual reaction in which UDP-Glc participates is the modification of the DNA of the T even bacteriophages (e.g. T_4) of *E. coli*, which contain the base 5-hydroxymethylcytosine instead of cytosine (10).

INTRODUCTION

The hydroxyl group of the base serves as an acceptor of glucose from UDP-Glc and attachment of the sugar results in the formation of a DNA which is protected from degradation by the nucleases of *E. coli*.

The discovery of sugar nucleotides led not only to the understanding of the biosynthesis of unusual monosaccharides and of complex saccharides, but also to the discovery in 1965 of a new type of activated sugars - the lipid-linked sugars. These are sugar derivatives linked to polyprenols by a mono- or diphosphate bridge. One example is bactoprenol (also known by a variety of other names, such as glycosyl carrier lipid) which in the form of its sugar diphospho derivative is an intermediate in the biosynthesis of bacterial lipopolysaccharides and peptidoglycan.

$$^{\ominus}O-\overset{O}{\underset{\underset{O^{\ominus}}{|}}{\overset{\|}{P}}}-O-CH_2-CH=\overset{CH_3}{\underset{|}{C}}-CH_2-(CH_2-CH=\overset{CH_3}{\underset{|}{C}}-CH_2)_9-CH_2-CH=\overset{CH_3}{\underset{|}{C}}-CH_3$$

Bactoprenyl phosphate

Recent work has demonstrated that similar compounds, the dolichol phosphates, participate in the biosynthesis of other microbial polysaccharides, and possibly also of glycoproteins produced by animal cells. The lipid-linked intermediates which are hydrophobic serve for the transport of activated sugars or oligosaccharides from the cytoplasm through the membrane to the cell surface or to the solution outside the cell. In connection with the study of the role of these intermediates in the biosynthesis of complex carbohydrates, new principles for the assembly of biological polymers have been discovered. Thus in the case of proteins or of simple polysaccharides (such as glycogen) synthesis proceeds by the addition of a single monomeric unit, in its activated form, to the growing polymer chain. However, with lipid-linked intermediates, a repeating oligosaccharide unit, for example a trisaccharide, is synthesized on the lipid carrier. The repeating unit is subsequently polymerized, and only then attached to a polymeric acceptor. This unusual mechanism, in which a polysaccharide is preassembled on a lipid carrier, operates in the biosynthesis of bacterial lipopolysaccharides, of peptidoglycan, and perhaps also of animal glycoproteins.

Studies of sugar nucleotides in relation to the biosynthesis of the bacterial cell wall peptidoglycan have led to the clarification of the mechanism of action of penicillin - still the most useful antibiotic. This has also heightened the interest in the structure of the bacterial cell wall. Speaking about the bacterial cell wall peptidoglycan, let me remind you that it is the substrate for lysozyme - the first enzyme whose three-dimensional structure was elucidated. This achievement of Phillips and his coworkers (11,12) was a landmark not only in the study of enzymes, but also in carbohydrate research.

NEW CONTROL MECHANISMS (13)

An enzyme for which a new type of control mechanism has been discovered is lactose synthetase, which catalyses the synthesis of the disaccharide lactose (Gal-β-(1→4)-Glc) from UDP-Gal and glucose, according to the equation:

$$\text{UDP-Gal} + \text{glucose} \longrightarrow \text{Gal-}\beta\text{-(1→4)-Glc} + \text{UDP}$$

The specificity of this enzyme is regulated by α-lactalbumin, which acts as a "specifier protein". In the absence of α-lactalbumin, the enzyme will not synthesize lactose but will transfer the galactose to N-acetylglucosamine to form instead the analogous disaccharide N-acetyllactosamine (Gal-β-(1→4)-GlcNAc) according to the following reaction:

$$\text{UDP-Gal} + \text{GlcNAc} \longrightarrow \text{Gal-}\beta\text{-(1→4)-GlcNAc} + \text{UDP}$$

It appears that lactose synthetase is normally involved in the biosynthesis of glycoproteins and can be changed into an enzyme with the specialized function of lactose synthesis. Presumably, the lactating mammary gland is the only tissue that forms lactose, because it is the only one capable of making α-lactalbumin. The discovery that α-lactalbumin, which has no catalytic activity of its own, is one of two proteins required for lactose synthetase activity is of unique biological significance. In addition to a better understanding of the mechanisms regulating the production of milk in the mammary gland, it has led to the discovery of other enzymes which consist of two different proteins.

GENETIC DISEASES

A completely different reason for the new wave of interest in carbohydrates stems from the fact that many of the genetic diseases of man, for which the molecular basis have been elucidated, are defects of carbohydrate metabolism, mostly of complex saccharides (Table 2). These include galactosemia (14), a defect in galactose metabolism, caused by the lack of a single enzyme - galactose 1-phosphate uridyl transferase. Galactosemia, like many other inherited disorders, causes mental retardation and often death at an early age. Other genetic diseases listed in Table 2 include the mucopolysaccharidoses - disorders of mucopolysaccharide

Table 2

Some genetic disorders of carbohydrate metabolism

Disorder	Defective enzyme or metabolic derangement
Aspartylglucosaminuria	N-acetylglucosaminyl-asparagine hydrolase
Fabry's disease	α-galactosidase
Fucosidosis	α-fucosidase
Galactosemia	Gal-1-P-uridyl transferase
Generalized gangliosidosis	absent β-galactosidase
Juvenile GM_1 gangliosidosis	deficient β-galactosidase
Juvenile GM_2 gangliosidosis	deficiency of hexosaminidase A
Glycogen storage disease Type 2	α-1-4-glucosidase
Hunter's disease	iduronate sulfatase
Hurler's disease	α-L-iduronidase
Mannosidosis	α-mannosidase ?
Sandhoff's disease	hexosaminidase A and B
Sanfilippo disease A	heparan sulfamidase
Sanfilippo disease B	α-N-acetylglucosaminidase
Tay-Sachs disease	hexosaminidase A

metabolism - such as Hurler's and Hunter's syndrome, and disorders of glycolipid metabolism, such as the Tay-Sachs disease, most common in East European (Ashkenazi) Jews. In most of these diseases the enzyme missing is normally involved in the degradation of complex saccharides.

CARBOHYDRATES AS DETERMINANTS OF BIOLOGICAL SPECIFICITY

Until recently, it was not appreciated that as a result of the possibility to form a large number of structures from a small number of monomers, nature can use sugars for the synthesis of highly specific compounds that will act as carriers of biological information. In other words, monosaccharides can serve as letters in a vocabulary of biological specificity, where the words are formed by variations in (a) the nature of the sugars present; (b) the type of linkages (α or β, $1 \rightarrow 2$, $1 \rightarrow 3$, etc.); and (c) the presence or absence of branch points. Indeed, we know now that the specificity of many natural polymers is written in terms of sugar residues, not of amino acids or nucleotides. This is an idea which is not entirely novel, but which has only recently become well established.

In the 1920's it was still believed that specific information carried in biopolymers involved only proteins. Between 1925 and 1937 O. T. Avery, at the Rockefeller

INTRODUCTION

Institute in New York, together with M. Heidelberger and W. F. Goebel, demonstrated that pure polysaccharides can carry specific immunological messages as antigens or haptens. Thus the highly purified Type III pneumococcus "specific soluble substance" was antigenic although it did not possess any of the properties of a polypeptide and was even found to be free of nitrogen. This substance was shown to be a polysaccharide, made up of cellobiuronic acid (a β-(1→4) linked disaccharide of glucuronic acid and glucose, GlcUA-β-(1→4)-Glc) linked through β-(1→3) linkages. The chemical basis of the antigenicity of polysaccharides was thoroughly clarified through the application of highly sophisticated end-group techniques developed by M. Heidelberger, E. A. Kabat, W. T. J. Morgan, O. Westphal and many others. At present it is well established that carbohydrates are ideally suited for the formation of specificity determinants that may be recognized by complementary structures, presumably proteins, on other cells or macromolecules.

Last, but not least, sugars have become the center of much interest because it has been shown that they serve as determinants of specificity on cell surfaces (15,16). The first indication for such a role came from the discovery in 1941 by G. K. Hirst in New York and by R. Hare in Toronto, that the influenza virus agglutinated erythrocytes. The molecular basis for this hemagglutination phenomenon was originally obscure. Mainly due to the efforts of A. Gottschalk in Australia (17) it was shown very clearly that the influenza virus binds to the red cell through sialic acid residues present on the cell surface. If the sialic acid is enzymically removed from the surface by neuraminidase, the influenza virus will no longer bind to the cell.

In 1952 it was shown that the specificity of the major blood types is determined by sugars. Thus, N-acetylgalactosamine is the immunodeterminant of blood type A, and galactose of blood type B. Interestingly, enzymic removal by specific glycosidases of α-linked N-acetylgalactosamine from type A erythrocytes, or of α-linked galactose from type B erythrocytes will convert both to type O erythrocytes. The latter conversion has been demonstrated by N. Harpaz in our Department using an α-galactosidase which he has purified from coffee beans (A5).

There are many additional examples of sugars as determinants of specificity on cell surfaces, and in shaping the social life of cells, from the interaction of bacteriophages with bacteria, through the sexual union in yeasts and the "homing of lymphocytes". The latter is an extremely interesting phenomenon, first demonstrated by B. M. Gesner and V. Ginsburg at the NIH in 1964 (A6). They found that rat lymphocytes, labeled with radioactive phosphate and injected into the rat through its tail, migrated to the spleen. However, if prior to injection the lymphocytes were treated with a specific glycosidase to remove $\underline{L}$-fucose from their surface, the lymphocytes migrated to the liver instead, as if the fucose on their surface dictated to the lymphocytes where to go. Work by G. Ashwell and A. G. Morell in the U.S. has now clearly established that sugars also serve as recognition signals for the survival and clearance of glycoproteins in blood serum.

Finally, alteration in sugar structure and architecture on cell surfaces appear to be intimately connected with the process of malignant transformation. These changes can be revealed by the use of plant agglutinins, now known as lectins, as has been shown by M. Burger in Princeton and Basel, by L. Sachs and M. Inbar at the Weizmann Institute, as well as by my own group (18).

I trust that what I have just told you has conveyed to you the feeling of where we stand at present in the study of carbohydrates, both simple and complex. It is also my hope that I have convinced you why this field is of such great importance, and why it is so exciting.

REFERENCES*

Books and review articles

1. An Introduction to the Chemistry of Carbohydrates,
   R. D. Guthrie and J. Honeyman (3rd ed.), Clarendon Press, Oxford, 1968, 144pp. (available also in paperback)
   The best concise description of the subject, a must for anybody who wants to take this course seriously.

2. Structural Carbohydrate Chemistry,
   E. G. V. Percival, revised by E. Percival (2nd ed.), J. Garnet Miller Ltd., London, 1962, 360pp.
   Somewhat outdated, but still a useful book.

3. Monosaccharide Chemistry,
   R. J. Ferrier and P. M. Collins, Penguin Books, England, 1972, 318pp. (in paperback)
   Treats the subject at a more advanced level than (1) and (2), using the current concepts of mechanistic organic chemistry and stereochemistry.

4. The composition and conformation of sugars in solution,
   S. J. Angyal, Angew. Chem. (Intern.Ed.) $\underline{8}$, 157-166 (1969).

5. The Shapes of Molecules, Carbohydrate Polymers,
   D. A. Rees, Oliver and Boyd, Edinburgh and London, 1967, 141pp. (paperback)
   A delightful short monograph, lucidly written, that conveys clearly the feel of modern research on carbohydrate polymers. Highly recommended.

6. Shapely polysaccharides,
   D. A. Rees, Biochem. J. $\underline{126}$, 257-273 (1972).

7. Distribution of amino sugars in microorganisms, plants and invertebrates,
   N. Sharon *in* The Amino Sugars (Eds. E. A. Balazs and R. W. Jeanloz), Academic Press, Vol.2A, 1965, pp.1-45.

8. Antibiotics containing sugars,
   S. Hannesian and T. H. Haskell *in* The Carbohydrates, Chemistry and Biochemistry (Eds. W. Pigman and D. Horton) Academic Press, 1970, Vol.2A, pp.139-211.

---

*This list does not include references to topics that will be discussed in more detail later in the course.

9. Two decades of research on the biosynthesis of saccharides,
    L. F. Leloir, Science 172, 1299-1303 (1971).
    Nobel Prize lecture of the man who contributed, perhaps more than anyone else, to the return of carbohydrates to the centre of the biochemical stage.

10. Virus Induced Enzymes,
    S. S. Cohen, Columbia University Press, New York, 1968, pp.47-50, 119-123.

11. The three-dimensional structure of an enzyme molecule,
    D. C. Phillips, Scient. Amer. 215 (5), 78-90 (1966).

12. Mechanism of lysozyme action,
    D. M. Chipman and N. Sharon, Science 165, 454-465 (1969).

13. The biological role of α-lactalbumin as a component of an enzyme requiring two proteins,
    K. E. Ebner, Acts. Chem. Res. 3 (2), 41-47 (1970).

14. Disorders of galactose metabolism,
    S. Segal *in* The Metabolic Basis of Inherited Disease (Eds. J. B. Stanbury, J. B. Wyngaarden and D. S. Fredrickson), 3rd ed., McGraw Hill, 1972, pp.174-195.

15. Structure and function of surface components of mammalian cells,
    V. Ginsburg and A. Kobata *in* Structure and Function of Biological Membranes (Ed. L. I. Rothfield), Academic Press, 1971, pp.439-459. *An excellent review.*

16. Surface Carbohydrates of the Eukaryotic Cell,
    G. M. W. Cook and R. W. Stoddart, Academic Press, 1973, 346pp.

17. The Chemistry and Biology of Sialic Acids and Related Substances,
    A. Gottschalk, Cambridge University Press, 1960, 115pp.
    Very readable account of the early history of the study of the sialic acids.

18. Lectins: cell-agglutinating and sugar-specific proteins,
    N. Sharon and H. Lis, Science 177, 949-959 (1972).

Specific articles

A1. The diaminohexose component of a polysaccharide isolated from *Bacillus subtilis*,
N. Sharon and R. W. Jeanloz, J. Biol. Chem. 235, 1-5 (1960).

A2. Structural studies of 4-acetamido-2-amino-2,4,6-trideoxy-$\underline{D}$-glucose (N-acetylbacillosamine), the N-acetyl-diamino sugar of *Bacillus licheniformis*,
U. Zehavi and N. Sharon, J. Biol. Chem. 248, 433-438 (1973)
   For the chemical synthesis of this and related compounds, see
A. Liav, J. Hildesheim, U. Zehavi and N. Sharon, Carbohyd. Res. 33, 217-227 (1974).

A3. The isolation of $\underline{D}$-erythro-$\underline{L}$-galacto-nonulose from the avocado, together with its synthesis and proof of structure through reduction to $\underline{D}$-arabino-$\underline{D}$-manno-nonitol and $\underline{D}$-arabino-$\underline{D}$-gluco-nonitol,
H. H. Sephton and N. K. Richtmyer, Carbohyd. Res. 2, 289-300 (1966).

A4. Biosynthesis of branched chain deoxysugars. IV. Isolation of cytidine diphosphate 6-deoxy-3-C-methyl-2-O-methyl-4-O-(O-methylglycolyl)-$\underline{L}$-aldohexopyranoside from *Azotobacter vinelandii*,
S. Okuda, N. Suzuki and S. Suzuki, J. Biol. Chem. 243, 6353-6360 (1968).

A5. Purification of coffee bean α-galactosidase by affinity chromatography,
N. Harpaz, H. M. Flowers and N. Sharon, Biochim. Biophys. Acta 341, 213-221 (1974).

A6. Effect of glycosidases on the fate of transfused lymphocytes
B. M. Gesner and V. Ginsburg, Proc. Natl. Acad. Sci. USA 52, 750-755 (1964).

2

GLYCOPROTEINS - I: GENERAL

Glycoproteins (1-5) are proteins to which carbohydrates are linked by glycosidic bonds. The carbohydrate moieties vary in size, from mono- or disaccharides to polysaccharides, and they are located at various positions on the polypeptide chain. To date we do not know of any glycoprotein that is built as a block copolymer, made up of alternating segments of peptides and oligosaccharides.

Glycoproteins are widely distributed in nature (Table 3). There is good reason to believe that most proteins are, in fact, glycoproteins. In other words, there are more proteins that contain covalently-bound carbohydrate in their molecule than proteins that are devoid of carbohydrate. Thus, of the over 60 proteins that have been isolated to date from human plasma, only serum albumin and prealbumin are not

Table 3

Distribution and function of some glycoproteins

Presumed function[a]	Source — Animal	Other
Structural	Collagen Mucopolysaccharides	Yeast cell wall Extensin (plant cell wall)
Food reserve	Casein Ovalbumin	Soybean 7S glycoprotein
Enzyme	Porcine ribonuclease B Porcine deoxyribonuclease Porcine amylase Acetylcholinesterase	Ficin Pineapple stem bromelain Taka-amylase Yeast invertase
Transport	Ceruloplasmin Haptoglobin Transferrin	
Hormone	Thyroglobulin Human chorionic gonadotropin Luteinizing hormone Follicle stimulating hormone Erythropoietin	
Protective	Fibrinogen Immunoglobulins Epithelial and submaxillary mucins Interferon	
Plasma and body fluids	Fetuin α_1-Acid glycoprotein (orosomucoid) α_2-Macroglobulin	
Toxins		Ricin Fungal phytotoxins
Unknown	Blood group substances Avidin (egg white)	

[a] Of intact molecules, not necessarily of carbohydrate portion

glycoproteins (Fig.3). All the other protein constituents

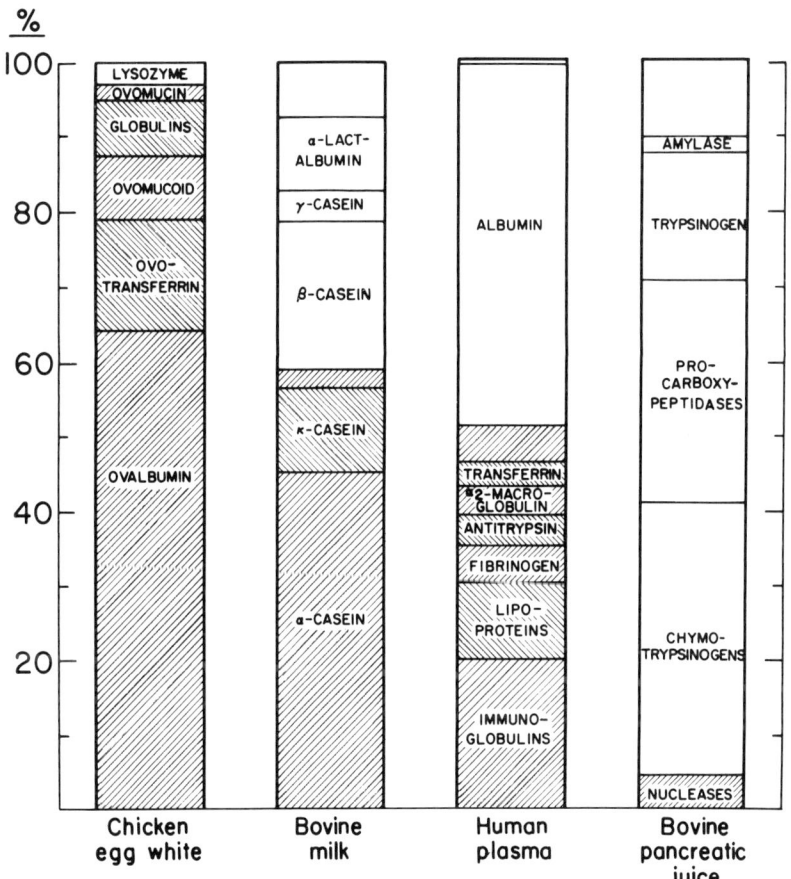

Figure 3 Quantitative distribution of proteins in animal fluids. The shaded areas represent glycoproteins. Modified from Winterburn and Phelps (6). (Numbers refer to % of total weight of protein in the fluid.)

of plasma - which by weight account for over 50 % of its total protein content - are glycoproteins. These include α_1-acid glycoprotein, fibrinogen, transferrin, ceruloplasmin, the immunoglobulins, and so on. In hen egg-white, all the proteins are glycoproteins, with the exception of lysozyme, which accounts for about 3% of the protein content of the egg-white. In contrast, in bovine pancreatic juice only about 5% of the proteins contain sugar. They include ribonuclease B, C and D, deoxyribonucleases and amylases. The collagens, which comprise a quarter to a third of the total protein of the human body, are all glycoproteins. Many hormones, e.g. follicle stimulating hormone, luteinizing hormone and thyrotropic hormone, are glycoproteins. An increasing number of enzymes have been shown to contain covalently bound sugar (7). In addition to those mentioned, the list of glycoprotein enzymes includes many others, such as fungal glucamylase, human parotid amylase, horseradish peroxidase, alkaline phosphatases, acetylcholine esterase, dopamine β-hydroxylase and cytoplasmic aspartic transaminase.

Recent literature abounds in reports of new glycoproteins with unusual properties. Of the many examples, I shall give you only a few. Human intestinal enzymes, specific for the hydrolysis of the disaccharides maltose and sucrose, have been shown to contain 30-40% carbohydrate by weight and to be resistant to proteolytic digestion by papain (A1). The major carbohydrates associated with these disaccharidases are fucose, galactose and hexosamines. Rather surprisingly, the purified enzymes demonstrated blood group antigenicity of great potency, as little as 8×10^{-9} to 1.5×10^{-7} M inhibiting red cell agglutination by the corresponding antisera. The enzymes displayed the same blood group specificity as the erythrocytes of the donor from whom they were isolated: enzymes from donors with blood type A possessed A activity and those from donors with blood type B had B activity. The blood group reactivity most probably resides in the oligosaccharide chain covalently linked to the disaccharidases. This is the first demonstration of blood group antigenicity associated with an enzymically active glycoprotein molecule.

Another very interesting glycoprotein is the material responsible for the sexual agglutination of cells from one type of yeast with those of another type. This specific sexual agglutination factor, isolated from *Hansenula wingei* type 5 haploid yeast cells (A2) is a glycoprotein with a molecular weight of about 9.6×10^5 which is composed of 85% carbohydrate, 10% protein and 5% phosphate. The protein part is unusual in that it contains 55% serine and 6-8% threonine. The isolated glycoprotein, as the type 5 cells from which it is obtained, will specifically agglutinate *H. wingei* type 21 cells. This ability is destroyed by digestion with pronase or with a bacterial exo-α-mannanase. Thus both the protein and the carbohydrate components contribute to the specific recognition and binding activities of the type 5 yeast sexual factor.

A third example which I would like to mention is from studies of the nervous system. For a long time it was believed that myelin - the white fatty substance forming a sheath around some nerve fibres - did not contain any glycoprotein. Rat brain myelin has, however, recently been shown to contain a glycoprotein component which constitutes about 1% of the total myelin protein (A3). Although the chemistry and function of this glycoprotein have not yet been established, it is believed to play an important role in the process of myelination. This discovery is also of interest with regard to the speculations about the possible role of glycoproteins in memory and nerve transmission.

Interferon, the antiviral agent produced by mammals, has very recently also been shown to be a glycoprotein (A4). Of great importance are the glycoproteins of cell membranes, which are attracting increasing attention. The best characterized of these is glycophorin (A5), the major glycoprotein of the human erythrocyte membrane, which I shall discuss in more detail in a later lecture.

Very unusual glycoproteins have been isolated from the sera of antarctic fish. They have been called "antifreeze glycoproteins", since they have the unique property of inhibiting the phase transition of water to ice and thus protect the fish from probable death by freezing (A6). These glycoproteins, of molecular weight 11,000 to 32,000, possess the simple repeating structure of Ala-Ala-Thr tripeptide, to which a disaccharide Gal-β-(1→3)-GalNAc is attached (Fig.4).

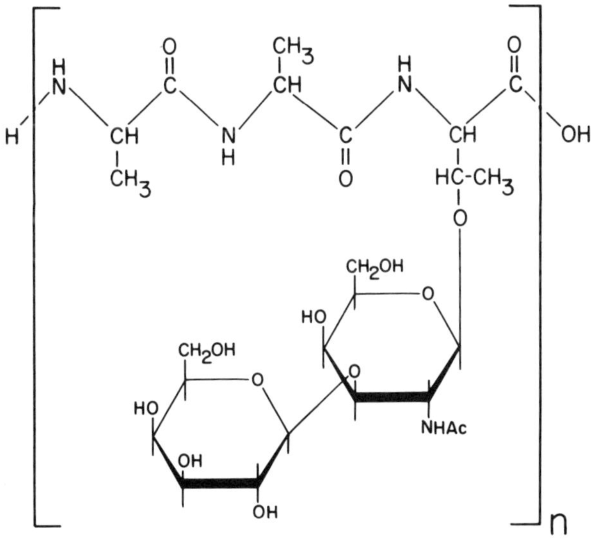

Figure 4 Structure of the antifreeze glycoproteins. The values of n differ for different glycoproteins isolated (e.g. n = 17, 28 or 35).

The disaccharide side chains are essential for the antifreeze activity of these glycoproteins, since their modification (e.g. by periodate oxidation) destroys this activity.

A group of compounds often classified as glycoproteins are the mucopolysaccharides, or proteoglycans, of connective tissue - e.g. the chondroitin sulfates, heparin and keratosulfate. They differ from the typical glycoproteins described before in that they contain long polysaccharide chains with a molecular weight of 20,000 or more (i.e. polymers made up of about 100 monosaccharide units). Another unique feature of mucopolysaccharides is that their carbohydrate side chains are composed of repeating disaccharide units. No repeating sequences are found in other glycoproteins where the saccharide side chains are much shorter - from a molecular weight of 180 (a monosaccharide) to about 3,000 (17 monosaccharide units). There are, however, carbohydrate structures that are common to many glycoproteins, such as the trisaccharide NANA-Gal-GlcNAc or the disaccharide NANA-GalNAc, but they do not occur in a repeating manner along the oligosaccharide side chain. Also, mucopolysaccharides usually contain uronic acids and sulfate groups, not found in typical glycoproteins. In recent years it has been shown, however, that brain glycoproteins are sulfated, as

GLYCOPROTEINS I

are gastric mucosa glycoproteins. It is possible that many other glycoproteins may be found, upon careful re-examination, to be sulfated.

Mucopolysaccharides were considered until the early 1960's to be pure carbohydrates, devoid of any protein. As we now know, this was because alkali was commonly used for their extraction from tissues; in their native form the polysaccharide chains of mucopolysaccharides are linked to proteins through linkages that are alkali-labile. Only when biochemists started to use proteolytic enzymes for the isolation of mucopolysaccharides from tissues did they observe that these compounds always contain covalently bound protein. In spite of the demonstration that mucopolysaccharides are glycoproteins, they continue to be considered as a separate group and will be discussed separately in this course.

For many years it has been known that glycoproteins cover the lining of the respiratory and intestinal tracts, are responsible for the viscosity of saliva and cervical mucus, and lubricate the eyeball in the eyesocket. These compounds, known as mucins, were the first - and until some 25 years ago the only - group of glycoproteins to be recognized as such (8). When other proteins were found to contain sugar they were believed to be impure and the sugar was assumed to be a contaminant. In fact, when a biochemist detected carbohydrate in a typical protein, i.e. one which was not mucin-like, he either did not pay attention to this finding, or made efforts to remove the sugar from the protein.

In many diseases changes are often observed in the levels of glycoproteins in body fluids and tissues (9). Although the reason for these changes is usually obscure, glycoproteins are attracting the attention of physicians. The increase in clinical interest in glycoproteins is also the result of the recognition that in a number of genetic defects (Table 2, page 25) accumulation of mucopolysaccharides or other glycoproteins occurs. In other diseases, the level of glycoproteins may be decreased. This is the case in Wilson's disease, which is characterized by defects in copper metabolism and a shortage of ceruloplasmin, a copper transport glycoprotein found in serum, which serves as a molecular link between copper and iron metabolism.

SUGAR CONSTITUENTS

Although over 100 different monosaccharides have been recognized in nature, only about a dozen have been found in glycoproteins (Table 4). Since many animal glycoproteins have now been carefully analysed, it is safe to assume that there is little likelihood of finding, at least in animal glycoproteins, sugars other than those given in the table.

Table 4

Monosaccharide constituents of glycoproteins^a

Hexoses	Galactose
	Mannose
	Glucose
Deoxyhexoses	L-Fucose
Hexosamines	N-Acetylglucosamine
	N-Acetylgalactosamine
Sialic acids	Acylneuraminic acids
Pentoses	Xylose
	L-Arabinose

^{a}In most glycoproteins, the configuration of the sugars present has not been proven conclusively. Typical sugar constituents of mucopolysaccharides are not listed.

Included in this list are xylose, found in the carbohydrate-peptide linking region of mucopolysaccharides, and L-arabinose, a typical constituent of glycoproteins of plants (but not of animals). I have not included in the table the uronic acids (D-glucuronic and its 5-epimer L-iduronic) which are components of the disaccharide repeating sequences of mucopolysaccharides, but are not found in other glycoproteins.

D-Glucuronic acid   L-Iduronic acid

The proportion of sugars in glycoproteins varies from 0.5% in some collagens, to 85% in blood group substances (Fig.5). It is possible that there are glycoproteins with an even higher content of sugar, since there is some evidence that glycogen contains small amounts of protein. If this protein is covalently bound, glycogen may be considered as a glycoprotein with over 99% sugar. The protein in this case may serve as the core, on which the polysaccharide chains of glycogen grow. It is possible, therefore, to view the carbohydrate spectrum of glycoproteins as extending essentially from 0 to 100%. Let us, however, not forget that there are "true" proteins, such as lysozyme, ribonuclease A and concanavalin A, that do not contain any sugar in their molecules.

As I said before, most proteins contain covalently bound sugar and are, therefore, glycoproteins. The reverse

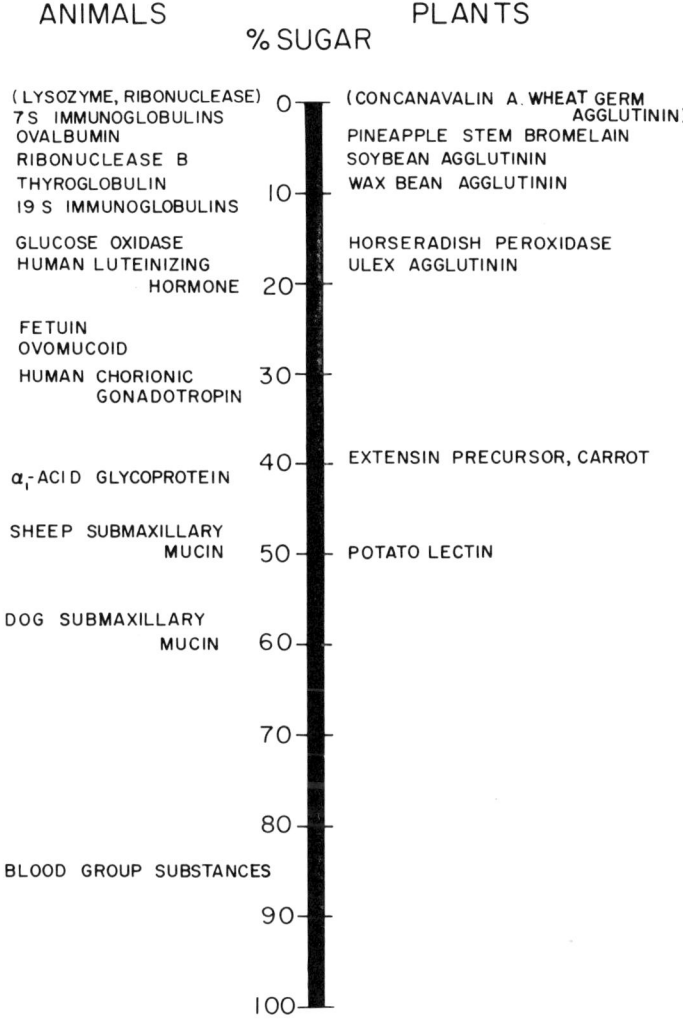

Figure 5 Carbohydrate content of glycoproteins from animals and plants.

is true for polysaccharides, especially those of animal tissues, many of which are now known to contain varying amounts of covalently bound protein.

REFERENCES

Books and review articles

1. Glycoproteins,
   N. Sharon, Scientific American 230 (5), 78-86 (1974).
   A suitable introduction to the subject.

2. Glcyoproteins,
   R. G. Spiro, Adv. Prot. Chem. 27, 349-467 (1973).
   A thorough and well presented coverage of the field. Highly recommended.

3. Glycoproteins,
   R. D. Marshall, Ann. Rev. Biochem. 41, 673-702 (1972).
   Interesting, with many provocative ideas.

4. Glycoproteins of higher plants,
   N. Sharon *in* Plant Carbohydrate Biochemistry (Ed. J. B. Pridham), Academic Press, 1974, pp.235-252.
   Deals with the distribution and properties of plant glycoproteins.

5. Glycoproteins,
   A. Gottschalk, Ed., 2 volumes, Elsevier, 1972, 1378pp.
   The most comprehensive coverage of the subject; a must for anybody working in the field.

6. The significance of glycosylated proteins,
   P. J. Winterburn and C. F. Phelps, Nature 236, 147-151 (1972).

7. Glycoenzymes: enzymes of glycoprotein structure,
   J. H. Pazur and N. N. Aronson Jr., Adv. Carb. Chem. Biochem. 27, 301-341 (1972).

8. Past and present concepts of glycoproteins
   A. Neuberger *in* Glycoproteins of Blood Cells and Plasma (Eds.G. A. Jamieson and T. J. Greenwalt), J. B. Lippincott Company, Philadelphia and Toronto, 1971, pp.1-15.
   A brief and interesting history of glycoprotein research.

9. Glycoproteins in disease
   R. J. Winzler *in* Glycoproteins of Blood Cells and Plasma (Eds.G. A. Jamieson and T. J. Greenwalt), J. B. Lippincott Company, Philadelphia and Toronto, 1971,pp.204-213.

Specific articles

A1. Blood group antigenicity of purified human intestinal disaccharidases,
J. J. Kelly and D. H. Alpers, J. Biol. Chem. 248, 8216-8221 (1973).

A2. Partial characterization of the sexual agglutination factor from *Hansenula wingei* Y-2340 type 5 cells,
P. H. Yen and C. E. Ballou, Biochemistry 13, 2428-2437 (1974).

A3. The enzymology of myelination,
R. O. Brady and R. H. Quarles, Molecular and Cellular Biochem. 2, 23-29 (1973).

A4. Interferon: evidence for its glycoprotein nature,
F. Dorner, M. Scriba and R. Weil, Proc. Natl. Acad. Sci. USA 70, 1981-1985 (1973).

A5. Molecular features of the major glycoprotein of the human erythrocyte membrane,
V. T. Marchesi, R. L. Jackson, J. P. Segrest and I. Kahane, Fed. Proc. 32, 1833-1837 (1973).

A6. A biological antifreeze,
R. E. Feeney, American Scientist 62, 712-719 (1974).

GLYCOPROTEINS - II: ISOLATION AND CHARACTERIZATION

For the isolation and purification of glycoproteins we usually employ the same methods as those used in the study of proteins devoid of carbohydrates. There are, however, some problems resulting from the presence of sugars in a protein molecule. Since sugars are polar compounds, glycoproteins tend to be more soluble in aqueous solutions and consequently it is not always possible to precipitate sugar-rich glycoproteins with typical protein precipitants such as ammonium sulfate. On the other hand, high solubility may in some cases be of help in the isolation of glycoproteins, such as α_1-acid glycoprotein and fetuin, which are among the serum glycoproteins richest in carbohydrate. These glycoproteins are soluble in trichloracetic and perchloric acid and can therefore be readily separated from the bulk of the other serum proteins and glycoproteins. The high negative charge imparted to certain glycoproteins by a large number of sialic acid, uronic acid or sulfate residues, has been used to advantage in separation procedures employing anion exchange chromatography, electrophoresis or precipitation by complex formation with quaternary ammonium salts.

In addition to changing the solubility and charge of proteins, sugars also affect their density. Because sugar molecules are significantly heavier (by about 15%) than those of amino acids, sugar-rich glycoproteins can be separated from proteins by density gradient centrifugation in cesium chloride.

During the last couple of years, a specific, mild and highly effective technique for the isolation of glycoproteins has been introduced, which is gaining much popularity. It is based on the ability of lectins to bind mono- and oligosaccharides in a specific and reversible manner. Lectins also interact with glycoproteins and precipitate them from solution. Such precipitates can be dissolved by the addition of the sugar(s) for which the lectin is specific. For example, concanavalin A will precipitate most glycoproteins from serum, leaving the albumin in solution. Dissolution of the precipitate is achieved by the addition of methyl α-mannoside or methyl α-glucoside for which concanavalin A is specific. A particularly useful extension of this technique involves separation of glycoproteins from proteins on columns of immobilized lectins, i.e. lectins linked to insoluble carriers such as the commercially available polymers Sephadex or Sepharose (A1). Glycoproteins that bind to such columns can be eluted by solutions of the specific sugar(s). Of the many examples appearing in the literature on the use of such columns, there is one that I find rather impressive. The glycoprotein enzyme dopamine β-hydroxylase, present in chromaffin vesicles of bovine adrenal glands, was isolated from lysates of these vesicles and obtained in pure form and high yield by a single passage through a column of concanavalin A-Sepharose (A2). Immobilized concanavalin A has also been used for the isolation of immunoglobulins and other serum proteins, of blood group substances and of many enzymes such as glucose oxidase. The availability of other lectins with different sugar specificities greatly increases the versatility of this method. Since lectins provide us with the only specific tool for the separation of proteins and glycoproteins, there is no doubt that their use for this purpose will continue to grow.

    A number of problems are often encountered in the

characterization of a purified glycoprotein. First and
foremost is the question of molecular homogeneity. Although
proteins proper are as a rule monodisperse, this is not
always the case with glycoproteins, which very often exhibit
molecular heterogeneity. An extreme example are the carbohydrate rich blood group substances, the molecular weight
of which ranges between $3 - 10 \times 10^5$. There are also great
difficulties in establishing the molecular weight of glycoproteins. Proteins containing a substantial amount of
carbohydrate behave in an anomalous manner during gel filtration and gel electrophoresis. While a linear relationship exists for most globular proteins between their elution
volumes on Sephadex gel columns and the logarithm of their
molecular weight, glycoproteins such as fetuin, ovomucoid
and thyroglobulin do not conform to the above relationship.
This appears to be due to a greater hydration in solution
brought about by the carbohydrate units, resulting in a
more expanded structure for glycoproteins than that for
proteins not containing carbohydrate. The use of gel filtration for the purpose of molecular weight determination
of glycoproteins is therefore precluded.

The finding that the migration of most proteins during
acrylamide gel electrophoresis in the presence of sodium

dodecyl sulfate (SDS) is inversely related to the logarithm of their molecular weights has served as the basis of a very effective tool for the characterization of a large number of proteins. Glycoproteins with high carbohydrate content have been found to migrate in this type of electrophoresis at rates slower than would be expected from their molecular mass. The low mobility of these carbohydrate-containing proteins was shown to be the result of their binding a smaller amount of SDS on a weight basis than standard proteins of the same mass. This method, too, can therefore not be used for reliable estimation of the molecular weight of glycoproteins.

PROBLEMS OF SUGAR ANALYSIS

The next question we usually ask is: how much sugar does the glycoprotein contain and what are its monosaccharide constituents? Although this may seem to be an easy problem to tackle, in fact it is a difficult one (1).

The total amount of sugar in glycoproteins is estimated colorimetrically by the use of old fashioned sugar reagents (A3,A4). One of the most useful of these is the phenol-sulfuric acid method, which measures the total content of neutral sugars. Some of the colorimetric methods are more specific; fucose, for example, can be estimated by the cysteine-sulfuric acid method of Dische and Shettles. However, most techniques for the quantitative determination of individual sugars in glycoproteins usually require the liberation of the monosaccharide from the glycosidic linkage. The reason for this is that, with the exception of the carboxyl group of sialic acids, the only groups present in the sugar moieties of glycoproteins which can be recognised by physical or chemical methods, without hydrolysis, are the hydroxyl groups and these are common to all types of sugar. This is in clear contrast to the side chains of the amino acid residues in proteins which are generally free, vary greatly in structure and reactivity and, as a result, can often be recognized and quantitatively determined by physical (e.g. optical) or chemical methods.

In order to liberate sugars from glycoproteins, hydrolysis by acid is at present almost invariably employed, as glycosidic bonds are in general stable to alkali. Hydrolysis by acid poses the most formidable problem encountered in the carbohydrate analysis of glycoproteins (1) and it may be instructive again to compare the situation with that found in the amino acid analysis of proteins in general. Apart from tryptophan, all amino acids commonly found in proteins survive the conditions of hydrolysis generally used

(6 M hydrochloric acid at 100-110°C, 24-72 h) either completely or with a limited and largely predictable amount of destruction, e.g. in the case of serine or threonine. By carrying out hydrolysis for varying periods of time, adequate correction can be made for losses with the two amino acids mentioned, and under the drastic conditions used all peptide linkages are split.

The situation is quite different in the analysis of the carbohydrate residues in glycoproteins, mainly because the stabilities to hot acid of the various types of monosaccharides encountered in glycoproteins vary greatly. The hexosamines are the group most resistant to destruction by acid, but they are less stable than the majority of amino acids; thus 5% of glucosamine or 10-14% of galactosamine is destroyed on heating in 4 M HCl at 100°C for 16 h. This destruction is not due to the effect of acid alone since it has been shown that if oxygen is excluded, the extent of destruction is greatly reduced. At the other extreme are the sialic acids which are rapidly destroyed on heating with dilute mineral acid; e.g. 0.01 M HCl for 30 min at 100°C causes 20% destruction of *N*-acetylneuraminic acid. Non-nitrogenous aldoses, such as mannose and galactose, occupy an intermediate position between these two extremes; e.g. 23%

of mannose was destroyed on heating for 5 h in 2 M HCl at 100°C.

For several reasons these values for the stabilities of pure sugars in hot mineral acid can be taken only as a rough guide in assessing the stabilities to be expected during hydrolysis of glycoproteins. In the first place, interaction occurs under the catalytic influence of acid between the free sugars and amino acids such as tryptophan, cysteine and methionine. Some correction can be made for the disappearance of the sugar by model experiments in which destruction is measured after heating suitable mixtures containing the sugar in question and amino acids similar to those present in the hydrolysate of the protein under investigation. Alternatively, a known amount of isotopically labelled sugar is added to the glycoprotein before hydrolysis, the sugar or a derivative thereof is isolated and the amount present in the glycoprotein is calculated from the dilution of the label. Secondly, the reducing group of a monosaccharide may react under the catalytic influence of acid with the primary or even secondary hydroxyl groups of another sugar molecule to give di- or oligosaccharides, a reaction called "acid reversion". Such a side reaction being bimolecular can be largely eliminated by carrying out the hydrolysis at a low carbohydrate concentration.

These are just a few of the difficulties encountered when attempting to release sugars from glycoproteins. At present it is impossible to say to what extent these and other sources of error affect the analytical results.

MECHANISM OF HYDROLYSIS OF GLYCOSIDES

In order to be able to choose the most suitable conditions for the hydrolysis of the different types of glycosidic bonds, it is necessary to consider in a more general manner the available facts and their interpretation (1,2). Acid hydrolysis of most glycosides is generally thought to proceed by a unimolecular mechanism and to involve preliminary protonation of the glycosidic oxygen as shown in Fig.6.

This protonation, leading to the conversion of compound I to II, is followed by a slow breakdown of the conjugate acid (compound II) to the cyclic carbonium ion (compound III). The latter, which is in the half-chair conformation (because of the partial double bond character of the C-1–O linkage) is attacked rapidly by water to give the free sugar (compound IV). However, even in solutions which are 0.1 M in glycoside, it is found that other glycoside molecules may compete with the solvent for the carbonium ion, and form

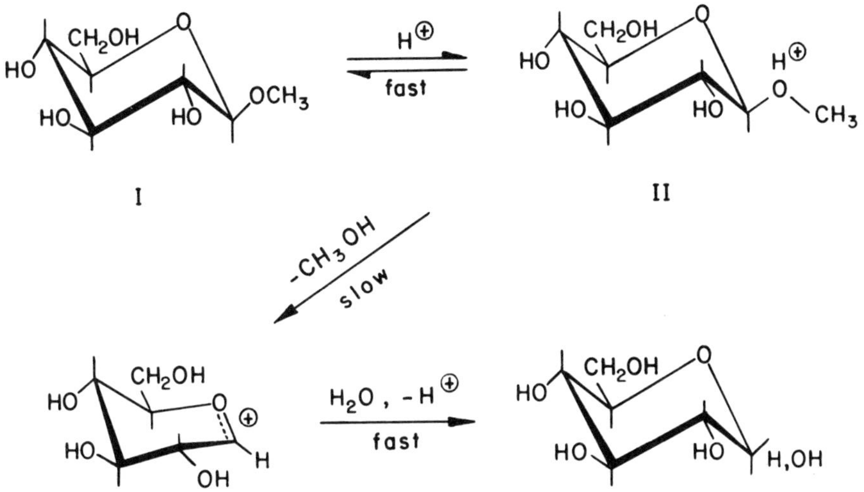

Figure 6 Mechanism of acid hydrolysis of glycosides. The glucose is represented here in its stable chair ($C1$) conformation (compounds I, II and IV). The carbonium ion intermediate (compound III) is in the half-chair conformation.

small amounts of disaccharides. Such a reaction is responsible for the "acid reversion" I mentioned a little earlier.

While the overall rate of hydrolysis will obviously depend on several factors, such as the character of the aglycon (in Fig.6 it is a methyl group), the conformation of the molecule and the size of the ring, the extent of protonation probably plays an important part. In particular the nature of the substituent on C-2 has been used to explain the differences in reactivity between the pyranosides of ordinary aldohexoses and of the 2-deoxyhexoses.

N-Acylneuraminic acid residues are removed from glycoproteins very readily, hydrolysis with 0.05 M H_2SO_4 at 80°C for 1 h being sufficient to complete the reaction. There are several reasons for this. In the first place neuraminic acid resembles other 2-deoxysugars, the glycosides of which have been reported to be hydrolyzed between 500 and 1000 times more readily than the corresponding derivatives of glucose. Secondly, a glycoside of N-acylneuraminic acid is a ketoside, which might be expected to be hydrolyzed more rapidly than an aldopyranoside. Thus fructopyranosides (also ketosides) are hydrolyzed by acid about 10,000 times faster than are the corresponding aldopyranosides. Finally N-acetylneuraminic acid is highly acidic (pK = 2.6) and therefore in 0.05 M H_2SO_4 about 2% of the acid is ionized. The un-ionized carboxyl group may be expected to inhibit protonation on the glycosidic oxygen in the same way as the carboxyalkyl group in glycine ethyl ester reduces the basicity of the amino group as compared with that found in ethylamine. On the other hand the ionization of the carboxyl group will largely abolish this effect.

Since L-fucose, a 6-deoxyhexose, is a constituent of many glycoproteins, it is worthwhile to note that fucosides appear to be about five times more readily hydrolyzed than the corresponding galactopyranosides or twenty-five times more readily than glucopyranosides. Fucose can be easily released from glycoproteins using mild conditions of acid hydrolysis, similar to those used for the release of sialic acid.

THE HYDROLYSIS BY ACID OF GLUCOSAMINIDES

In 1938, Moggridge and Neuberger suggested that the great resistance of α- or β-methylglucosaminide to hydrolysis by acid was caused by the presence of a positive charge in close proximity to the glycosidic linkage. It was also proposed that methyl N-acetylglucosaminide could be hydrolyzed by acid along two pathways (Fig. 7). The rate constants k_1, k_2 and k_3 were all approximately of the same order of magnitude but the rates of hydrolysis of α- or β-methylglucosaminide (k_4) were about 250 times lower.

Almost all glycoproteins contain N-acetylhexosamine residues and it is therefore not surprising that similar problems are encountered in the hydrolysis of all natural

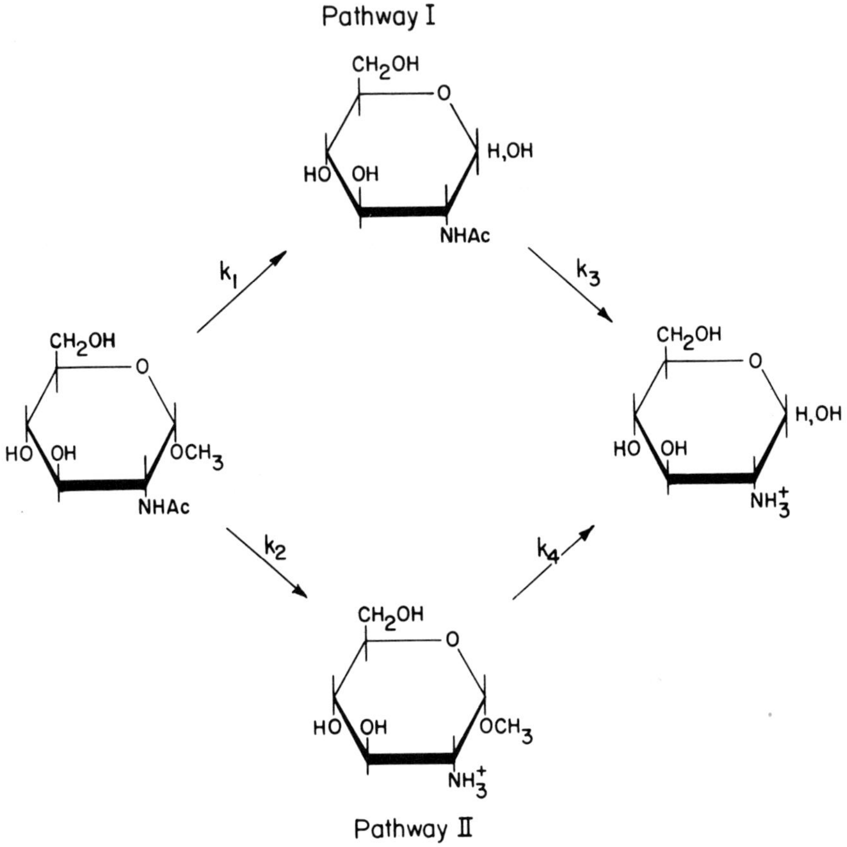

Figure 7 Pathways of acid hydrolysis of the N-acetyl-glucosaminyl linkage. The example given is that of methyl α-N-acetylglucosaminide.

products containing hexosamine residues. Furthermore, any hexose residue to which N-acetylhexosamine is glycosidically linked might be incompletely liberated if hydrolysis is carried out under the conditions usually employed for the release of neutral sugars (e.g. 1 M H_2SO_4, 6 h, 100°C).

Hydrolysis by pathway II will not only be slow, but the recovery of glucosamine is likely to be far from quantitative, since k_4 might not be very much greater than the rate of destruction of glucosamine under the vigorous conditions necessary for its complete liberation. Conditions must therefore be chosen in such a way that as large a proportion as possible of the glucosamine present should be liberated by the favourable pathway I. It appears that the use of high concentrations of acid (3 - 4 M) and temperatures of about 100°C satisfy the requirements. Concentrations of acid higher than 4 M and temperatures above 100°C lead to marked destruction of glucosamine and thus the conditions of hydrolysis recommended (4 M HCl at 100°C for 3 h) represent a compromise between the maximal liberation of glucosamine and its minimal destruction.

Special problems are also encountered in the hydrolysis of mucopolysaccharides. Total acidic hydrolysis of these compounds yields a mixture of the constituent monosaccharides, although extensive decarboxylation of the uronic acid residues usually results. Recently, the use of new reagents (e.g. potassium borohydride) has permitted reduction of carboxyl groups (after esterification) in the mucopolysaccharide, and identification of the uronic acid as the parent hexose following acid hydrolysis. The major product of limited hydrolysis of mucopolysaccharides is usually the constituent aldobiuronic acid, i.e. a disaccharide consisting of a hexosamine (or hexose) and a hexuronic acid. Such a product is formed because the carboxyl function of the alternate uronic acid residues shields their glycosidic oxygen from attack by hydrogen ions.

Once the sugars have been released by the hydrolytic methods discussed above, they can be identified and estimated by a variety of procedures. Identification is done by chromatography on paper or on thin layers, by electrophoresis on paper and, increasingly, by gas liquid chromatography. The latter method (A5), which is becoming the technique of choice, requires the conversion of the monosaccharides or of their reduction products, the alditols, into volatile derivatives, most commonly either by acetylation or by trimethylsilylation.

The above methods can also be used for quantitative analysis, with gas liquid chromatography displacing all other techniques. In fact, using gas liquid chromatography it is now possible to identify and quantitate, in a single procedure, all the monosaccharides present in a glycoprotein. Still widely employed are enzymic assays, measurement of reducing power and colorimetric techniques (A3,A4,A6). Among the enzymes used for sugar analysis, the most popular ones are glucose oxidase (highly specific) and galactose oxidase (reacts also with galactosamine). Measurements of reducing power and colorimetric reactions for sugars are, unfortunately, of low specificity. For this reason it is important to apply them only after the sugars in glycoprotein hydrolysates have been separated from each other (and from the amino acids), preferably by chromatographic means. Glucosamine and galactosamine are determined either by the Elson-Morgan color reaction or on an amino acid analyser.

Acid hydrolysis in aqueous solution may now be replaced by methanolysis, which causes less destruction of the sugars. The resultant methyl glycosides are then converted to volatile derivatives which can be identified and estimated quantitatively by gas liquid chromatography.

REFERENCES

Reviews

1. Qualitative and quantitative analysis of the component sugars,
   R. D. Marshall and A. Neuberger, *in* Glycoproteins (Ed. A. Gottschalk), 2nd ed., part A, Elsevier, 1972, pp.224-299.

2. Acid-catalyzed hydrolysis of glycosides,
   J. N. BeMiller, Advan. Carb. Chem. $\underline{22}$, 25-108 (1967).

Specific articles

A1. Group-specific separation of glycoproteins,
    T. Kristiansen, Meth. Enzymol. $\underline{34}$, 331-341(1974).

A2. The interaction of dopamine-β-hydroxylase with concanavalin A and its use in enzyme purification,
    R. A. Rush, P. E. Thomas, S. H. Kindler and S. Udenfriend, Biochem. Biophys. Res. Comm. $\underline{57}$, 1301-1305 (1974).

A3. Color reactions of carbohydrates,
    Z. Dische, Meth. Carb. Chem. $\underline{1}$, 477-514 (1962).

A4. New colorimetric methods of sugar analysis,
    G. Ashwell, Meth. Enzymol. $\underline{8}$, 85-95 (1966).

A5. The determination of carbohydrate in biological materials by gas-liquid chromatography,
    J. Clamp, T. Bhatti and R. E. Chambers, Meth. Biochem. Anal. $\underline{19}$, 229-344 (1971).

A6. Analysis of sugars found in glycoproteins,
    R. G. Spiro, Meth. Enzymol. $\underline{8}$, 3-26 (1966).

4

CARBOHYDRATE-PEPTIDE LINKAGES

Although all glycoproteins conform to the same general structural pattern in that they consist of polypeptide chains to which carbohydrate moieties are attached, they differ in many of their properties. Thus, the molecular size of glycoproteins ranges from 15,000 to over a million. They differ not only in the relative proportion of sugar, and the types of the latter, but also in the number of side chains present.

Some glycoproteins are of small molecular size and contain only one carbohydrate chain per molecule. Prominent examples are ovalbumin, the major protein of hen egg white (molecular weight 45,000), and the enzyme ribonuclease B secreted from bovine pancreas of the ox (molecular weight 14,700). At the other extreme we have proteins such as sheep submaxillary mucin (molecular weight of about one million) which contain some 800 saccharide chains in each molecule. In sheep submaxillary mucin, each side chain is composed of two sugar residues, and the protein has quite a dense sugar distribution - one disaccharide unit for every six to seven amino acids. It follows from the above that the frequency of occurrence of oligosaccharides along the polypeptide chain varies markedly from one protein to another (Table 5).

The most characteristic feature of glycoproteins is the presence of a covalent linkage between carbohydrate and protein (2). Of the twenty or so amino acids from which proteins are made up, only five have been found to participate in the formation of linkages with carbohydrates. Three of these - L-asparagine, L-serine and L-threonine - are common protein constituents, while the other two - 5-hydroxy-L-lysine and 4-hydroxy-L-proline - occur only rarely. There is no obvious structural similarity between these amino acids, although certain relationships exist between the triple-base codons for some of them. Each codon for L-asparagine (AAU and AAC) can give rise, by single base substitutions, to codons for L-serine (AGU and AGC, respectively), L-threonine (ACU and ACC), and L-lysine (AAA and AAG). There are no codons for hydroxylysine and hydroxyproline, since these amino acids are formed by hydroxylation of lysine and proline, respectively. This hydroxylation occurs after completion of the

Table 5

Variations in spacing and number of carbohydrate units in glycoproteins[a]

Glycoprotein	Number of units per molecule	Spacing of units (amino acids/unit)
1. Serum glycoproteins		
α_1-Acid glycoprotein	4	51
Fetuin	6	60
Human haptoglobin	13	113
Human α_2-macroglobulin	31	209
Calf thyroglobulin	19	296
Human transferrin	2	375
Human IgG	2	776
2. Pancreatic glycoproteins		
Bovine ribonuclease B	1	124
Deoxyribonuclease	1	270
3. Mucins		
Ovine submaxillary mucin	800	6
Porcine submaxillary mucin (A^+)	~500	8
4. Collagens		
Bovine glomerular basement membrane	–	58
Rabbit corneal collagen	19	173
Calf skin collagen	8	435
Rat skin collagen	4	770
Rabbit scleral collagen	3	1,000

[a]Modified from Spiro (1).

polypeptide chains into which lysine and proline have been incorporated.

Table 6 lists the amino acids and sugars that participate in the formation of carbohydrate-peptide linkages. These linkages are of two chemical types: N-glycosidic and

Table 6

Carbohydrate-peptide linkages in glycoproteins

Linkage	Stability to alkali	Occurrence
N-Glycosidic		
N-Acetylglucosaminyl-asparagine	±	Many animal and plant glycoproteins
O-Glycosidic		
N-Acetylgalactosaminyl-serine or threonine	–	Mucins, blood group substances, membrane glycoproteins, immunoglobulins, fetuin, antifreeze glycoproteins
Xylosyl-serine	–	Mucopolysaccharides (proteoglycans)
Galactosyl-serine	–	Earthworm cuticle collagen, plant cell walls
Mannosyl-serine	–	Yeast cell wall mannans, yeast invertase, *Aspergillus niger* glucoamylase
Mannosyl-threonine	–	Earthworm cuticle collagen, fungal phytotoxic glycopeptides
Galactosyl-hydroxylysine	+	Collagens, basement membrane
L-Arabinosyl-hydroxyproline	+	Plant cell walls and glycoproteins

O-glycosidic; they differ markedly in their chemical properties, in particular their stability to acid and base.

Glycoproteins may, to a large extent, be subdivided according to the nature of their carbohydrate-peptide linkages. Such a classification is not rigidly applicable since the same glycoprotein molecule may contain two different types of prosthetic groups linked by different carbohydrate-protein linkages. The occurrence of both N-glycosidic and O-glycosidic carbohydrate-peptide linkages in the same molecule has been observed in immunoglobulins, fetuin, human chorionic gonadotropin, erythrocyte membrane glycoproteins and the glomerular basement membrane.

In glycoproteins which possess only one type of linkage, there may be a number of saccharide chains which are not necessarily identical, and which may differ in size and structure. For example, calf thyroglobulin with a molecular weight of 670,000 contains two classes of carbohydrate units, both of which are linked to the protein via N-acetylglucosaminyl-asparagine linkages: (a) five relatively short chains, with a molecular weight of about 2,000 and composed of mannose and N-acetylglucosamine only and (b) 14 longer chains, molecular weight 3,000, comprised of mannose, N-acetylglucosamine, N-acetylgalactosamine, $\underline{L}$-fucose and sialic acid.

THE N-ACETYLGLUCOSAMINYL-ASPARAGINE LINKING GROUP (3)

The first carbohydrate-peptide linkage to be identified was that between N-acetylglucosamine and asparagine (GlcNAc-Asn or Asn-GlcNAc). The bonded sugar and amino acid were isolated from ovalbumin in 1963 after a long, sustained effort by A. Neuberger and R. D. Marshall in London; this linkage was identified in the same year by L. W. Cunningham in the U.S., by I. Yamashina in Japan and by V. P. Bogdanov in the U.S.S.R.

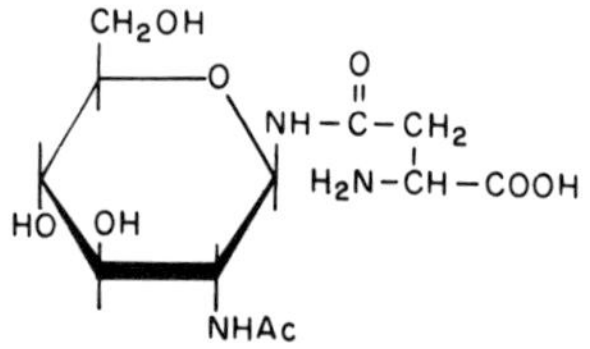

Figure 8 Structure of N-acetylglucosaminyl-asparagine (2-acetamido-1-($\underline{L}$-β-aspartamido)-1,2-dideoxy-β-$\underline{\underline{D}}$-glucose).

In this compound (Fig.8), the anomeric carbon atom of N-acetylglucosamine is linked glycosidically by a β-linkage to the amide group of asparagine. Alternatively, the compound can be viewed as a condensation product of the β-glycosylamine of N-acetylglucosamine (or more accurately 2-acetamido-2-deoxy-β-$\underline{\underline{D}}$-glucopyranosylamine) and the β-carboxyl of aspartic acid. One way of forming such a linkage is by acylating the β-amine of the sugar with the β-carboxyl of aspartic acid. In fact, GlcNAc-Asn was synthesized some ten years ago in a number of laboratories (A1,A2) by condensation of α-benzyl-N-benzyloxy-carbonyl-aspartate with the β-amine of 3,4,6-tri-O-acetyl-N-

acetylglucosamine, followed by removal of the protecting groups (Fig.9). The synthetic product was identical in all

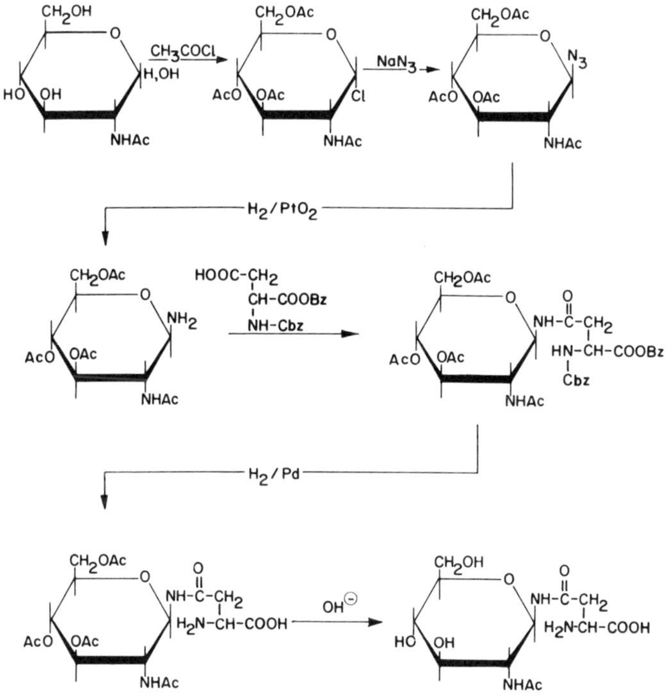

Figure 9 The synthesis of N-acetylglucosaminyl-asparagine

respects to GlcNAc-Asn isolated from ovalbumin, thus establishing unequivocally the structure of the linking group.

The N-acylglycosylamine linkage is relatively stable to mild acids. In 2 M hydrochloric acid at 100°C, GlcNAc-Asn decomposes with a half-life of about 17 minutes, with the release of aspartic acid and ammonia. The linkage is likewise relatively stable under mild alkaline conditions, ammonia being released by 0.2 M sodium hydroxide at 100°C, with a half-life of about 100 minutes. Alkaline cleavage of GlcNAc-Asn in the presence of $NaBH_4$ converts the N-acetylglucosamine into N-acetylglucosaminitol. When such treatment (1 M NaOH - 1 M $NaBH_4$, 4-6 hrs, 100°C) is applied to glycopeptides with GlcNAc-Asn linkages, complete cleavage of the linkages occurs, the N-acetylglucosamine involved in the linkage is converted into N-acetylglucosaminitol and the products can be used to establish the pattern of substitution of the N-acetylglucosamine residue (A3) (see pp.105-106).

PROTECTION OF SUGARS AGAINST ALKALI BY BOROHYDRIDE

We have seen here an example of the use of sodium borohydride to cleave alkali-labile sugar linkages. The role of $NaBH_4$ is to convert the reducing group (aldehyde or hemiacetal group) of the sugar released by the alkali, into an alcoholic group. This conversion protects the free sugar from destruction by alkali.

The fact that free sugars are sensitive to alkali has been known for many years; the initial step of the action of dilute alkali, such as lime water, on a reducing monosaccharide (e.g. glucose) is a partial transformation to the C-2 epimeric aldose and the corresponding ketose. In this reaction, known since 1895 as the Lobry de Bruyn transformation, glucose gives an equilibrium mixture with fructose and mannose, through the postulated intermediate 1,2-enol. In addition, dilute alkali may cause degradation of reducing sugars resulting in the production of such breakdown products as glyceraldehyde. Rearrangements will also take place in concentrated alkali causing the formation of carboxylic derivatives known as saccharinic acids. None of these reactions will occur with polyalcohols, so that conversion of the reducing group of the sugar into an alcoholic function will stabilize the sugar against degradation by alkali.

Radioactive tritium-labelled $NaBH_4$ (or KBH_4) is often used, which greatly facilitates the detection, identification and quantitative determination of minute amounts of the reduction products (A4). Reduction with NaB^3H_4 (known also as $NaBT_4$) is particularly useful, since this compound can be obtained with very high specific radioactivity - perhaps the highest of any compound available on the market.

O-GLYCOSIDIC LINKAGES TO SERINE AND THREONINE (3,4)

The O-glycosidic linkages to serine and to threonine

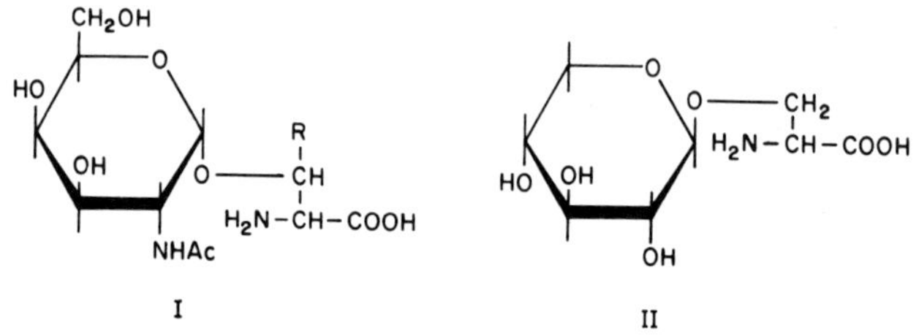

Figure 10 O-glycosidic linkages to serine (R=H) or threonine (R=CH$_3$). Note that whereas the linkage to N-acetylgalactosamine is of the α-configuration (I), that to xylose is β (II).

(Fig.10), such as GalNAc-Ser, GalNAc-Thr, or Xyl-Ser, are very labile to alkali, and will be completely cleaved even by 0.05-0.1 M sodium hydroxide at room temperature within 24 hours.

This type of cleavage is known as the β-elimination reaction, an alkali catalyzed bimolecular reaction which may be represented as follows:

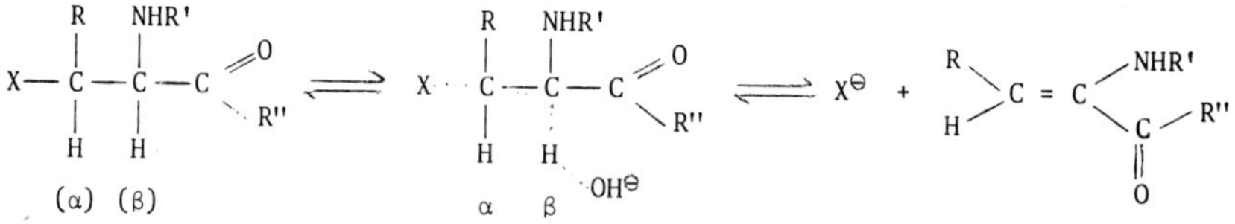

Transition state

where X is the GalNAc or Xyl group, R is H (for serine) or CH$_3$ (for threonine), and R' and R" are other amino acids of the peptide chain. The group X must be strongly electron-attracting. It should be pointed out that in the notation generally used to describe this type of reaction, the α and β carbon atoms are the reverse of those usually employed in amino acid chemistry.

Such alkaline cleavage converts serine and threonine to their unsaturated derivatives 2-aminoacrylic acid (dehydroalanine) and 2-aminocrotonic acid, respectively.

$$\begin{array}{c} CH_2OH \\ | \\ H_2N-CH-COOH \end{array} \longrightarrow \begin{array}{c} CH_2 \\ \| \\ H_2N-C-COOH \end{array}$$

      serine   2-aminoacrylic acid

These unsaturated compounds have a characteristic absorption in the ultraviolet and their formation can therefore be readily detected. The conversion into 2-aminoacrylic and 2-aminocrotonic acids reduces the level of the hydroxyamino acids in the original glycoprotein. Therefore, a reduction in the levels of serine and threonine after alkali treatment of a glycoprotein may be taken as an indication of the involvement of these amino acids in a carbohydrate-protein linkage.

If alkali treatment is carried out in the presence of borohydride, the unsaturated amino acid derivatives are converted to alanine of the DL configuration and 2-aminobutyric acid (also DL), respectively; the latter compounds are stable to acid hydrolysis and can be determined by amino acid analysis.

Alkali treatment under reducing conditions converts the peptide-linked monosaccharide to the corresponding sugar alcohol, and therefore serves to identify the sugar of the linkage region. Use of radioactive $NaBH_4$ may, of course, facilitate the identification of the newly formed amino acids and sugar alcohols.

The β-elimination reaction, and in particular the conversion of the serine and threonine into alanine and 2-aminobutyric acid, respectively, are rarely quantitative. It is important to establish the proper conditions for the reaction with each glycoprotein in order to approach complete release of the bound carbohydrate, and quantitative conversion of the β-eliminated amino acids. When the β-elimination reaction is carried out in the presence of sulfite (A5), the sugar-linked hydroxyamino acids are converted almost quantitatively into cysteic acid or 2-amino-3-sulfonyl butyric acid, which can be separated and determined either in the amino acid analyser or by gas liquid chromatography of their trimethylsilyl derivatives.

Proteins or peptides with dehydroalanine may, under suitable conditions, be cleaved at the peptide bond next to the amino group of this amino acid. The product will be a pyruvoyl peptide, from which the pyruvic acid can be released, identified and quantitated by chemical or enzymic techniques.

Linkage analysis by alkali treatment is fraught with certain pitfalls, since a prerequisite for facile β-elimination is the presence of substituents on both the amino and carboxyl groups of the hydroxyamino acid (serine and threonine). This means that the reaction only proceeds smoothly if the hydroxyamino acids do not occupy terminal position in the peptide. Therefore, failure to observe destruction of serine and threonine on alkali-treatment of a glycoprotein cannot be taken as definitive proof of the absence of such carbohydrate-peptide linkages. It is advisable to establish the nature of a carbohydrate-peptide linkage not only by the effects of alkali treatment, but also by actual isolation and characterization of a linkage fragment containing the respective monosaccharide and amino acid.

All that I have just said is, of course, true also for the xylosyl-serine linkage. And as I told you in a previous lecture, the alkali lability of this linkage was the reason why, for a long time, mucopolysaccharides were believed to be "pure" polysaccharides, and not glycoproteins (or proteoglycans).

THE GALACTOSYL-HYDROXYLYSINE LINKAGE (3)

Another type of O-glycosidic linkage found in glycoproteins is between galactose and 5-hydroxy-L-lysine (Gal-Hyl) (Fig.11).

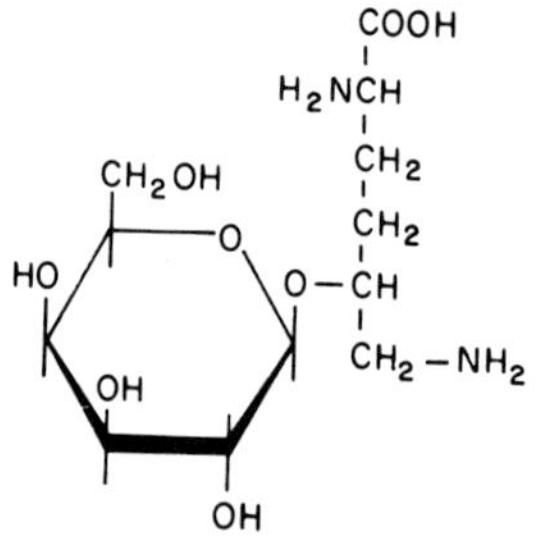

Figure 11 The O-glycosidic linkage between galactose and hydroxylysine

This linkage, first identified by W. T. Butler and L. W. Cunningham in 1966, is confined almost exclusively to the collagens, including the basement membrane of bovine kidney glomeruli. It is of the β-D-configuration and is very stable under alkaline conditions, much more so than peptide linkages under the same conditions. Compounds in which galactose or the disaccharide Glc-(1→2)-Gal are linked to the 5-hydroxyl group of hydroxylysine have been isolated in high yield from alkaline hydrolysates of collagen glycopeptides. The conditions of hydrolysis were rather drastic: 2 M sodium hydroxide at 90-105°C for 16-20 hrs. The galactosyl-hydroxylysine and glucosyl-galactosyl-hydroxylysine released can be separated and estimated quantitatively on the amino acid analyser; both compounds have free amino groups, and thus give a positive ninhydrin reaction. The galactosyl-hydroxylysine linkage is also stable under acidic conditions (e.g. 0.05 M sulfuric acid at 100°C for 28 hrs). This has been ascribed to the presence of the positively charged ε-amino group, which has the effect of stabilizing the glycosidic linkage to a considerable degree. Indeed, the rate of release of galactose by acid from Gal-Hyl glycopeptides is markedly increased by N-acetylation of the ε-amino group. The synthesis of Gal-Hyl has not been accomplished, the limiting factor being the availability of optically pure 5-hydroxy-L-lysine. There seems, however, little reason to suspect any difficulty in achieving the necessary reactions.

THE ARABINOSYL-HYDROXYPROLINE LINKAGE (5)

The last type of linkage, that between 4-hydroxy-L-proline and L-arabinose, has been recognized more recently (Fig.12). It is also O-glycosidic and alkali stable. To

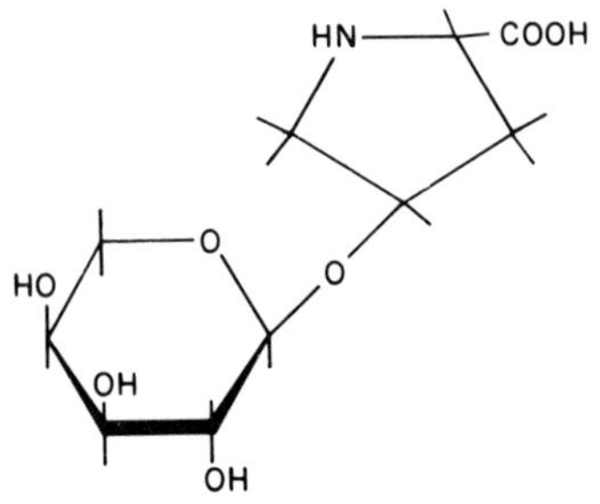

Figure 12   The O-glycosidic linkage between L-arabinose and 4-hydroxy-L-proline. Although the anomeric configuration of this linkage is not known, it is depicted here arbitrarily as being of the β form.

date it has been identified only in plant proteins, mainly in extensin, a hydroxyproline-rich plant cell wall protein which, upon hydrolysis, yields oligosaccharides of L-arabinose linked to 4-hydroxy-L-proline. In addition, extensin contains another type of linkage, that between galactose and serine. Apparently arabinosyl-hydroxyproline also serves as the carbohydrate-peptide linking group in the potato lectin, recently shown to contain 50% arabinose and 16% hydroxyproline (A6).

UNUSUAL LINKAGES

In addition to the linkages just discussed, there are

reports in the literature describing other, novel types of carbohydrate-peptide bonds. Some of these, such as the O-glycosidic linkage between mannose and serine or threonine, found in yeast invertase or collagen, respectively, are listed in Table 6.

The recent finding of glycopeptides containing thioglycosidic linkages is of some interest, particularly since one of the glycopeptides, with a glucose trisaccharide linked to cysteine, was isolated from erythrocyte membranes. The amino acid sequence of this unusual glycopeptide resembled that of the other glycopeptide with a thioglycosidic linkage which was isolated from urine. It was suggested that the urinary glycopeptide, which contained a galactose disaccharide, originated from the kidney membrane. The thioglycosidic linkage between the oligosaccharides and the SH group of cysteine is hydrolyzed in alkali through a β-elimination reaction, analogously to the O-glycosidic linkages to serine and threonine.

The existence of a novel type of glycosidic bond, formed between the phenolic hydroxyl group of tyrosine and the glycosidic hydroxyl of sialic acid, was recently suggested (A7). Such a carbohydrate-peptide linking group was reported to be present in hen ovomucoid, although the evidence for this structure was not very convincing.

REFERENCES

Reviews

1. Glycoproteins,
   R. G. Spiro, Ann. Rev. Biochem. $\underline{39}$, 599-638 (1970).

2. Aspects of the structure and metabolism of glycoproteins,
   R. D. Marshall and A. Neuberger, Adv. Carb. Chem. Biochem. $\underline{25}$, 407-478 (1970).
   Contains a particularly good discussion of the various carbohydrate-peptide linkages.

3. Carbohydrate-peptide linkages in glycoproteins and methods for their elucidation,
   A. Neuberger, A. Gottschalk, R. D. Marshall and R. G. Spiro, *in* Glycoproteins (Ed. A. Gottschalk), 2nd ed., Elsevier, 1972, pp.450-490.

4. Carbohydrate-peptide linkages in proteoglycans of animal, plant and bacterial origin,
   U. Lindahl and L. Rodén, *in* Glycoproteins (Ed. A. Gottschalk), 2nd ed., Elsevier, 1972, pp.491-517.

5. The glycopeptide linkages of extensin: O-$\underline{D}$-galactosyl serine and O-$\underline{L}$-arabinosyl hydroxyproline,
   D. T. A. Lamport, *in* Biogenesis of Plant Cell Wall Polysaccharides (Ed. F. Loewus), Academic Press, 1973, pp.149-164.

Specific articles

A1. Carbohydrates in protein. VIII. The isolation of 2-acetamido-1-($\underline{L}$-β-aspartamido)-1,2-dideoxy-β-$\underline{D}$-glucose from hen's egg albumin,
   R. D. Marshall and A. Neuberger, Biochemistry $\underline{3}$, 1596-1600 (1964).

A2. The synthesis of a glucosamine-asparagine compound. Benzyl N^2-carbobenzyloxy-N-(2-acetamido-3,4,6-tri-O-acetyl-2-deoxy-β-$\underline{D}$-glucopyranosyl)-$\underline{L}$-asparaginate,
   C. H. Bolton and R. W. Jeanloz, J. Org. Chem. $\underline{28}$, 3228-3230 (1963).

A3. A common structural unit in asparagine-oligosaccharides of several glycoproteins from different sources,
Y. C. Lee and J. R. Scocca, J. Biol. Chem. 247, 5753-5758 (1972).

A4. Quantitative determination of reducing sugars, oligosaccharides, and glycoproteins with [^{3}H]borohydride,
C. McLean, D. A. Werner and D. Aminoff, Analyt. Biochem. 55, 72-84 (1973).

A5. β Elimination and sulfite addition as a means of localization and identification of substituted seryl and threonyl residues in proteins and proteoglycans,
D. L. Simpson, J. Hranisavljevic and E. A. Davidson, Biochemistry 11, 1849-1856 (1972).

A6. The purification and properties of the lectin from potato tubers, a hydroxyproline-containing glycoprotein,
A. K. Allen and A. Neuberger, Biochem. J. 135, 307-314 (1973).

A7. Studies on tyrosine environments of chicken ovomucoid. The environment of the most-exposed tyrosine,
M. A. Krysteva, I. N. Mancheva and I. D. Dobrev, Eur. J. Biochem. 40, 155-161 (1973).

5

GLYCOPEPTIDES - I: ISOLATION AND CHARACTERIZATION

Our information on the nature and properties of carbohydrate-peptide linkages is based on two approaches. One of these we have seen in the case of the alkali-labile O-glycosyl serine or threonine linkages, where most studies have been carried out on intact glycoproteins. This approach is, however, of limited application.

The best way to identify a carbohydrate-peptide linkage is by its isolation from a glycoprotein. For this purpose it is necessary first to digest the glycoprotein with proteolytic enzymes and from the hydrolysate to isolate glycopeptides in which intact carbohydrate side chains are linked to short peptide segments, or sometimes to one amino acid only.

Proteolytic digestion is most commonly done with pronase, but other proteases, singly or in sequence, also serve this purpose. The digest is fractionated by gel filtration on Sephadex (Fig.13) or Bio-Gel, which readily separates the free amino acids and short peptides from the oligosaccharide-peptides (or oligosaccharide-amino acid compounds), since the latter are usually of a larger molecular size.

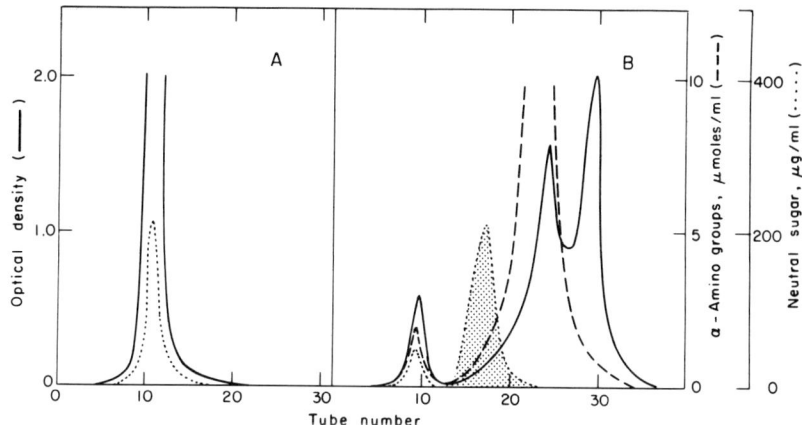

Figure 13 Gel filtration on Sephadex G-50 of purified soybean agglutinin (A) and of soybean agglutinin after exhaustive digestion with pronase (B) (from reference A1). Shaded area represents the glycopeptide fraction. Optical density measured at 280 nm.

I would, however, like to emphasize that there is no general way for the purification of such glycopeptides to homogeneity: methods such as ion-exchange chromatography and electrophoresis are being used for this purpose, but it is inherently almost impossible to obtain homogeneous glycopeptides. This difficulty stems, to a large extent, from the microheterogeneity of the carbohydrate side chains of glycoproteins, a property which we shall discuss in greater detail later on. Suffice it to say, at this point, that while in a purified and homogeneous protein all molecules are identical in composition and sequence, this is not the case for glycoproteins. The carbohydrate side chains of a single glycoprotein, even when carefully isolated and purified from a genetically pure organism, are not identical in all the molecules of the glycoprotein. Thus, the single glycopeptide (or asparginyl-carbohydrate) of ovalbumin from a single egg of a pure-bred hen, is in fact a mixture of compounds which differ in composition and structure.

It is possible, however, to obtain glycopeptides which, to a first approximation, can be considered as homogeneous. If the purified glycopeptide contains a single amino acid - as we have found in our laboratory with a glycopeptide isolated from soybean agglutinin which contained only aspartic acid (A1) - the identity of the linking amino acid is obvious. But even when there are two, three or four amino acids, it may not be too difficult to identify the one to which the sugar is linked, especially if only one of the amino acids has a side chain which can potentially form linkages with sugars. Unequivocal identification of the linking group may, however, require removal of the extra amino acids, which is not always easy to achieve.

Before continuing, let me remind you that many glycoproteins are resistant to proteolytic enzymes. When dealing with such glycoproteins, it is necessary to modify them so that they become susceptible to proteolysis. Thus, proteins rich in sialic acid lose their resistance to proteases after removal of this sugar, either by enzymic treatment with neuraminidase or by mild acid hydrolysis. The carbohydrate side chains obtained after such treatment are, of course, no longer intact, a fact to be kept in mind when studying the isolated glycopeptides.

USE OF GLYCOSIDASES

To further identify the carbohydrate-peptide linking group, the isolated oligosaccharide-amino acid compound is degraded either by partial acid hydrolysis or by suitable enzymes. An example of the former method is the isolation of GlcNAc-Asn from the asparaginyl carbohydrate of ovalbumin (A2). Since the N-glycosidic linkage in this compound is more stable to acid hydrolysis than the O-glycosidic linkages between its sugar residues, controlled acid hydrolysis cleaves most of the latter linkages without affecting the former.

Enzymic digestion of the glycopeptide is a much more useful technique, since it is applicable not only for the isolation of the carbohydrate-peptide linking group, but also for elucidating the structure of the intact glycopeptide. The enzymes used for this purpose are exoglycosidases that remove sugar residues singly from the non-reducing end of oligo- or polysaccharides. Unfortunately, the situation with respect to the availability of glycosidases has, until very recently, been much less satisfactory than with proteases. Enzymes of the latter class, with a wide range of specificities, are available on the market, and are not expensive. In contrast, glycosidases are only now becoming commercially available, and they are expensive. The best way to obtain purified glycosidases is still to prepare them in the laboratory. A point to remember in this connection is the great care that should be taken to use purified and highly specific glycosidases in structural studies of glycoproteins and glycopeptides. Thus, in studies of the configuration of mannosidic linkages, it is essential to use α-mannosidase preparations that are devoid of β-mannosidase since β-mannosidic linkages are liable to be split by trace amounts of the β-specific enzyme. This, in fact, was the reason why all mannose residues in glycoproteins were, until very recently, believed to be α-linked, which proved to be incorrect when highly purified α-mannosidase became available.

Table 7

Glycosidases for the study of the carbohydrate side chains of glycoproteins

Enzyme	Source	Commercially available
Neuraminidase	*Clostridium perfringens*	+
	Diplococcus pneumoniae	
	Vibrio cholerae	+
β-N-Acetylglucosaminidase	Beef kidney	
	Diplococcus pneumoniae	
	Jack bean	+
	Pig epididymis	
	Proteus vulgaris	
	Turbo cornutus	+
α-Mannosidase	*Charonia lampas*	
	Jack bean	+
	Proteus vulgaris	
β-Mannosidase	Jack bean	
	Hen oviduct	
β-Galactosidase	*Aspergillus niger*	
	Charonia lampas	+
	Diplococcus pneumoniae	
	Escherichia coli	+
α-Galactosidase	Coffee beans	+
	Charonia lampas	+
α-$\underline{L}$-Fucosidase	Beef kidney	+
	Charonia lampas	
α-N-Acetylgalactos-aminidase	*Lumbricus terestris*	
	Porcine liver	

A list of some glycosidases used in structural studies of glycopeptides is given in Table 7. With such enzymes, singly or in mixture, it is possible to peel off the sugars of a glycopeptide one after the other, and to end up with a compound composed of a monosaccharide and a single amino acid - the carbohydrate-peptide linking group. This was, in fact, what we obtained from the asparaginyl-carbohydrate of soybean agglutinin upon incubation of this glycopeptide with a mixture of mannosidase and β-N-acetylglucosaminidase from Jack bean meal (A3). The final product was identified as GlcNAc-Asn, showing that the linking group in soybean agglutinin, a plant glycoprotein, is identical to that found in many animal glycoproteins.

Most glycosidases, though they readily remove sugars from short glycopeptides, will not remove sugars from intact glycoproteins. To take soybean agglutinin again as an example: this glycoprotein consists of 4 identical subunits, each with a molecular weight of 30,000 and each of which carries a carbohydrate side chain of the approximate structure Man_5-GlcNAc-Man_4-GlcNAc. No mannose, however, could be released by its prolonged incubation with high concentrations of α-mannosidase, although such removal was readily achieved upon incubation of the purified glycopeptide with small amounts of the enzyme.

Certain other glycosidases will remove sugar residues from glycoproteins without any difficulty. Thus, neuraminidase releases sialic acid from intact glycoproteins and β-galactosidase will then remove galactose from these desialated glycoproteins. The latter enzyme also acts on intact glycoproteins with terminal non-reducing β-galactose residues.

A new endoglycosidase, which will certainly prove of great value in studies of glycoproteins, has recently been isolated from cultures of *Streptomyces griseus* (A4). This enzyme removes carbohydrate side chains composed of N-acetylglucosamine and mannose not only from glycopeptides but also from intact glycoproteins, such as ovalbumin or ribonuclease B. The enzyme acts by cleaving the bond between the GlcNAc-Asn and the rest of the carbohydrate side chain.

CHEMICAL TECHNIQUES: METHYLATION AND SMITH DEGRADATION

In addition to enzymes, chemical techniques are also extensively used for structural studies of the carbohydrate units of glycoproteins. One of the most widely used approaches has been permethylation of free hydroxyl groups, followed by acid cleavage, and identification of the partially methylated monosaccharide units. An important advance in this methodology was the introduction in 1964 of a procedure by S. I. Hakomori, then at Sendai University, Japan, that utilizes methyl iodide and dimethylsufinyl carbanion, the latter being a more powerful nucleophile than the bases previously used in methylation. This reagent leads to rapid and complete methylation of all free hydroxyl groups as well as N-methylation of the acetamido group in hexosamine residues, without loss of N-acetyl groups. Techniques have been elaborated by B. Lindberg and his associates in Stockholm for the hydrolysis of such permethylated derivatives, conversion of the substituted monosaccharides into partially methylated alditol acetates, and identification of the latter through gas liquid chromatography or gas liquid chromatography and mass spectrometry (1). Methylation is usually done on isolated glycopeptides and only rarely on intact glycoproteins.

Another very useful approach is periodate oxidation, in particular in its modified form known as the Smith degradation. This method, developed in the 1950's by F. Smith at the University of Minnesota (2), involves periodate oxidation, reduction with borohydride and mild acid hydrolysis of the polyalcohol formed. Quantitation of the periodate uptake and of the formic acid released, together with detailed analysis of the degradation products, gives information on sequences of sugar residues in the original saccharide.

The principle of the method will become clearer after we review briefly some features of the periodate oxidation reaction. Oxidation of a simple glycoside such as methyl α-D-glucopyranoside with periodic acid or its salts (e.g. sodium periodate) yields a dialdehyde and formic acid (one mole per mole of glycoside). In the process, two molecules of periodate are used up.

The same result will be obtained if the glycoside is substituted at the 6 position. If the substitution is at the 2- or 4-hydroxyl, only one mole of periodate will be consumed per mole of glycoside, a dialdehyde will be formed, but no formic acid will be produced.

Sugars which are either 3-O-substituted, or 2,4-O-disubstituted, so that they do not contain free vicinal hydroxyls, are resistant to periodate oxidation.

The dialdehydes obtained by periodate oxidation readily form cyclic products of different structure, as shown in the example given below:

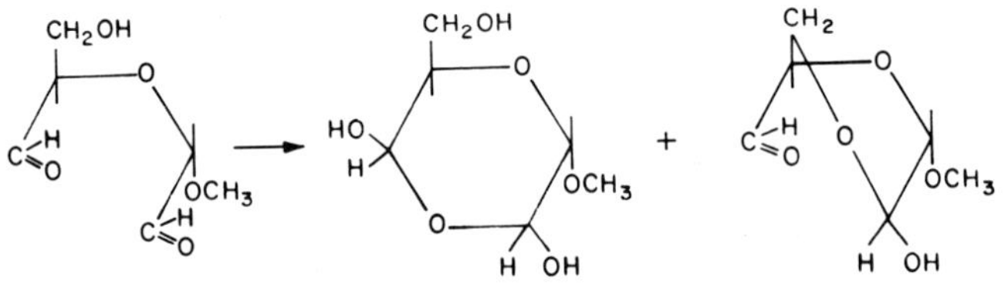

These cyclic acetals are relatively acid stable, and it is difficult to establish their structure. In the past, the dialdehydes were oxidized by bromine to the corresponding dibasic acids and these acids have been used to identify the structure of the glycosides. It was, however, recognized that if the aldehyde groups are reduced to the corresponding alcohols, the products obtained being true acetals are sensitive to acid (Fig.14). In fact, the rate of hydrolysis is so fast (up to 10^5 times faster than that of methyl α-glucoside), that it is possible to achieve virtually complete hydrolysis of these acetals under conditions when glycopyranosidic linkages are affected insignificantly, if at all.

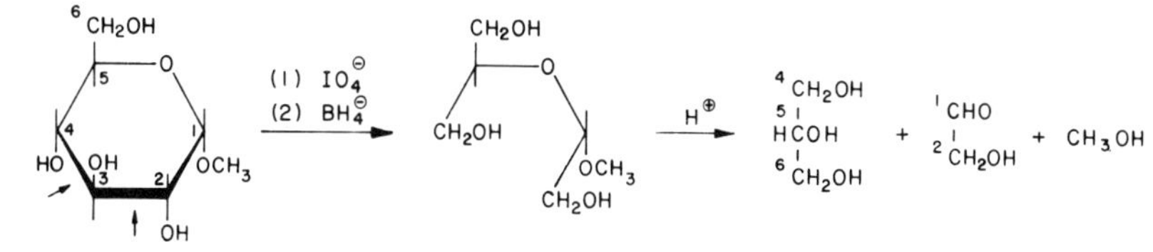

Figure 14 Products of acid hydrolysis of the alcohols obtained upon periodate oxidation and sodium borohydride reduction of methyl α-D-glucopyranoside (I) and of methyl 4-O-methyl-α-D-glucopyranoside (II). The carbons are numbered for clarity.

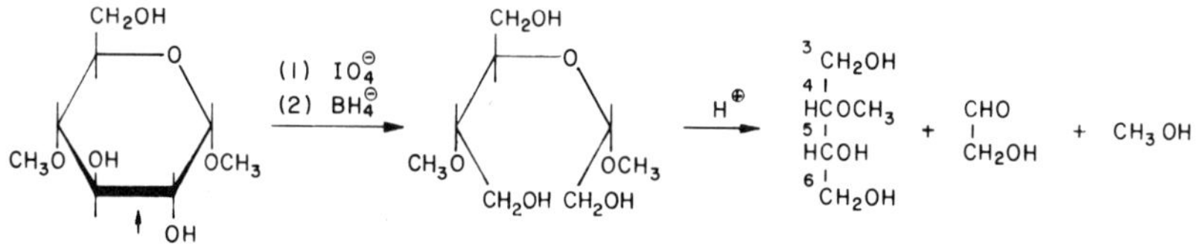

Characterization of the hydrolysis products provides important information on the nature of the sugar residue which has been degraded (Fig.14). Thus, Smith degradation of methyl α-glucopyranoside gives glycerol from C6-C5-C4, glycolaldehyde from C1-C2 and methanol from the aglycon, in addition to formic acid from C-3; methyl 4-O-methyl-α-glucopyranoside affords instead of glycerol 2-O-methyl-erythritol and no formic acid.

When the Smith degradation is applied to polysaccharides, a mixture of products is obtained, which may include alcohols (glycerol, erythritol), aldehydes, and glycosides of mono-, di- or higher oligosaccharides, the latter originating from sugar residues which are resistant to oxidation. Detailed analysis of the products throws light on the fine structure of the parent polysaccharides.

As an example, let us see what will happen when we degrade the polysaccharide nigeran by the Smith technique.

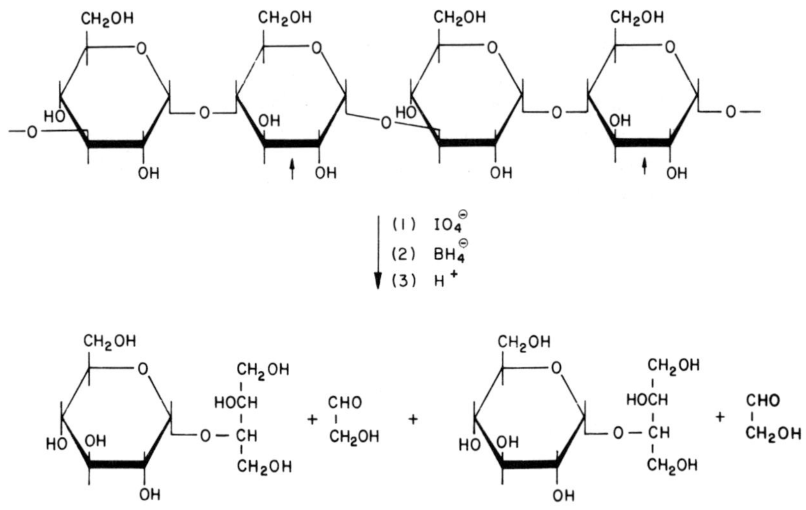

Figure 15 Smith degradation of nigeran. The bonds cleaved by periodate are marked by arrows.

Nigeran is a glucan in which the glucose units are joined by alternating α-(1→3) and α-(1→4) linkages. Upon periodate treatment, the 3-O-linked glucose residues will remain intact, while the 4-O-linked ones will be oxidized (Fig.15). Reduction of the oxidized polysaccharide, followed by mild acid hydrolysis, yields glycolaldehyde and 2-O-α-glucopyranosyl-erythritol. These compounds can be readily identified, the latter one by hydrolysis with stronger acid to glucose and erythritol.

As another example, let us examine the products of the Smith degradation of a hypothetical oligosaccharide containing an amino sugar (Fig.16). In this oxidation, 3 moles of per-

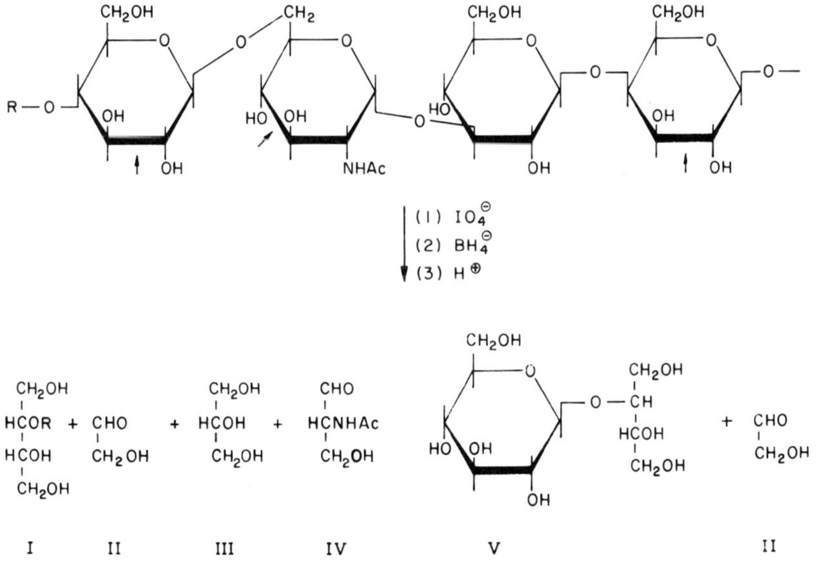

Figure 16 Products of Smith degradation of Glc-β-(1→6)-GlcNAc-α-(1→3)-Glc-β-(1→4)-Gal. The bonds cleaved by periodate are marked by arrows. The products obtained are erythritol substituted in position 2 (I), glycolaldehyde (II), glycerol (III), 2-acetamido-3-hydroxypropanol (IV) and β-glycosylthreitol (V).

iodate will be consumed. One of the compounds produced in the example given, 2-acetamido-3-hydroxypropanol (Fig.16, compound IV) was isolated as a product of the Smith degradation of the ovalbumin glycopeptides (see p.106).

REFERENCES

Review articles

1. Gas-liquid chromatography and mass spectrometry in methylation analysis of polysaccharides,
    H. Björndal, C. G. Hellerqvist, B. Lindberg and S. Svensson, Angew.Chem. Intern. Ed. $\underline{9}$, 610-619 (1970).

2. Controlled degradation of polysaccharides by periodate oxidation, reduction and hydrolysis,
    I. J. Goldstein, G. W. Hay, B. A. Lewis and F. Smith, Methods Carb. Chem. $\underline{5}$, 361-370 (1965).

Specific articles

A1. Soybean hemagglutinin, a plant glycoprotein. I. Isolation of a glycopeptide,
    H. Lis, N. Sharon and E. Katchalski, J.Biol.Chem. $\underline{241}$, 684-689 (1966).

A2. The properties of a carbohydrate-amino acid complex from ovalbumin,
    I. Yamashina and M. Makino, J.Biochem.(Tokyo) $\underline{51}$, 359-364 (1962).

A3. Identification of the carbohydrate-protein linking group in soybean hemagglutinin,
    H. Lis, N. Sharon and E. Katchalski, Biochim. Biophys. Acta $\underline{192}$, 364-366 (1969).

A4. The release of intact oligosaccharides from specific glycoproteins by endo-β-N-acetylglucosaminidase H,
    A. L. Tarentino, T. H. Plummer, Jr., and F. Maley, J.Biol.Chem. $\underline{249}$, 818-824 (1974).

6

GLYCOPEPTIDES - II: STRUCTURAL FEATURES

MICROHETEROGENEITY

The early preparations of the glycopeptide from ovalbumin contained five moles of mannose and three of N-acetylglucosamine per mole of aspartic acid. No other amino acids were present in the best preparations of this glycopeptide, which is therefore more correctly referred to as "asparaginyl-carbohydrate". Such preparations were, however, shown to be heterogeneous by L. W. Cunningham and his coworkers in 1965 (1). Subsequently, the asparaginyl-carbohydrate was separated into five fractions (A to E in Fig.17), in which the ratio of mannose to N-acetylglucosamine varied from 7:5 to 5:2 (A1). A more recent study (2) gave somewhat different results (Table 8). Heterogeneity of the type described is known as "microheterogeneity" (1,2).

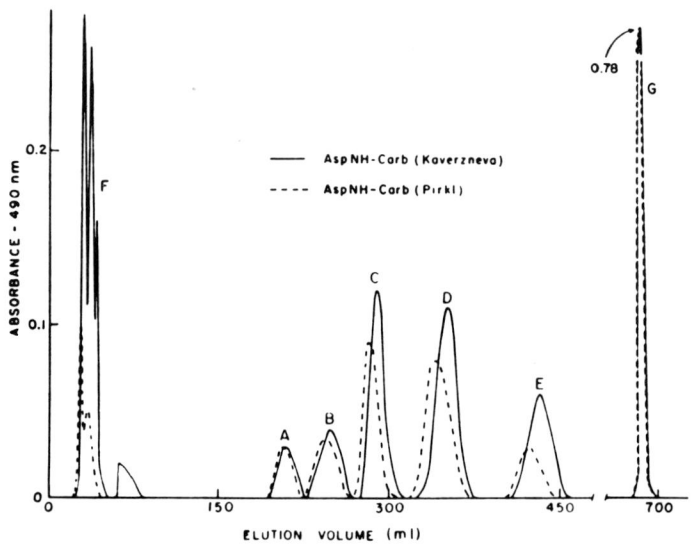

Figure 17 Fractionation of the asparaginyl-carbohydrate from two sources of chicken ovalbumin on a column of AG-50W x 2 resin, 0.9 x 150 cm. Elution was accomplished with sodium acetate buffer, pH 2.6. Absorbance at 490 nm is a measure of the content mannose, as estimated with the phenol-sulfuric acid method. (From reference A1). Fractions C and D have recently been further separated, into two fractions each (R. Montgomery, private communication).

Because of microheterogeneity, it is impossible to write a single chemical structure for the carbohydrate moiety of ovalbumin; it is best described by the following formula:

$$(\text{GlcNAc})_{0 \text{ or } 1 \text{ or } 2} - \text{Man-GlcNAc-GlcNAc} \overset{\mid}{\underset{\mid}{\text{Asn}}}$$

$$\text{Man}_{0 \text{ or } 1} - (\text{Man})_3$$

$$\text{Man}(\text{GlcNAc})_{0 \text{ or } 1}$$

Table 8

Composition of the asparaginyl-carbohydrate from chicken ovalbumin (2)

Comp- onent	Composition[a]		Molecular weight[b]	Sugar residues hydrolyzed by	
	Mannose	N-acetyl- glucosamine		α-manno- sidase	β-N-acetyl- glucosamin- idase
A	6.01	5.00	2142	1.0	2.9
B	5.10	5.08	2022	0.0	3.0
C	5.92	4.05	1970	1.1	2.0
D	5.94	2.18	1634	4.7	0.4
E	5.01	1.96	1387	3.8	0.0

[a] Moles per mole of aspartic acid.
[b] Based on aspartic acid content.

A structure which is closely related to ovalbumin is found in bovine ribonuclease B. The entire amino acid sequence and conformation of this protein are known, and it was shown by T. H. Plummer and C. H. W. Hirs in 1964 that all of the carbohydrate is attached in a single prosthetic group to asparagine 34. Although microheterogeneity has apparently not been demonstrated on a single preparation, ribonuclease B isolated in different laboratories by different procedures has been reported to have variable ratios of mannose to N-acetylglucosamine. The probable structure of the asparaginyl-carbohydrate from ribonuclease B is given below, where

$n = 1$ to 5 (1).

$$\begin{array}{c}\text{Asn}\\|\\(\text{Man})_n\text{-Man-GlcNAc-GlcNAc}\\|\\(\text{Man})_{5-n}\end{array}$$

Microheterogeneity has been observed in carbohydrate side chains of other glycoproteins as well. As a result, it is often very difficult to establish the structure of the carbohydrate moieties of glycoproteins and it is not surprising that cases are known where two investigators, studying apparently the same compound, proposed different structures for it. It should also be appreciated that we do not have sensitive criteria for assessing the degree of homogeneity of glycopeptide preparations at present. It appears that chromatography on Dowex 50 at very low ionic strength is the most efficient method developed so far for the separation of closely related asparaginyl-carbohydrates and for establishing their homogeneity.

The microheterogeneity of the carbohydrate side chains of glycoproteins might arise either by an unfinished biosynthetic sequence (which we shall discuss in detail later on) or by a post-biosynthetic degradation of completed carbohydrate chains. At present, no distinction between these two alternatives, biosynthetic microheterogeneity and degradative microheterogeneity, can be made.

Homogeneity is, of course, a rule for proteins, where microheterogeneity seems to be rarely encountered. It is common, however, not only in the carbohydrate chains of glycoproteins, but in polysaccharides in general, both simple and complex.

COMMON STRUCTURAL FEATURES

Relatively simple carbohydrate chains of the type found in ovalbumin and bovine ribonuclease B, which are composed solely of mannose and N-acetylglucosamine and are linked to the protein via asparagine, are found in a number of other glycoproteins, for example in bovine deoxyribonuclease A, in α-amylase from *Aspergillus oryzae* and soybean agglutinin. Another closely related class of glycopeptides contains the same two sugars and the same type of linkage, together with galactose, N-acetylgalactosamine, $\underline{\text{L}}$-fucose and sialic acid. In such complex saccharide chains, found for example in fetuin, in thyroglobulin, and in α_1-acid glycoprotein, the mannose and glucosamine are arranged near the protein in a so-called core, whereas the other sugars, together with additional residues of N-acetylglucosamine, are found in a branched outer region (Fig.18).

```
NANA            NANA      NANA                              NANA
 │α2→3           │α2→3     │α2→3                             │α2→3
 ↓               ↓         ↓                                 ↓
Gal             Gal       Gal                               Gal
 │β1→4           │β1→4     │β1→4                             │β1→3
 ↓               ↓         ↓                                 ↓
GlcNAc          GlcNAc    GlcNAc               NANAα(2→6)GalNAc
 │β1→2           │β1→3(4)  │β1→2(4,6)                         │α
 ↓               ↓         ↓                                 ↓
Manα(1→2(6))Manα(1→3)Man(1→4)GlcNAcβ(1→4)GlcNAcβAsn    Ser(Thr)
```

Figure 18 Structures proposed for the asparagine-linked and serine (threonine)-linked carbohydrate units of fetuin. (The latter also occur in the form of a trisaccharide without N-acetylneuraminic acid (NANA) linked to N-acetylgalactosamine). (A2,A3).

The sialic acid and L-fucose are always peripherally located, and their reducing group is engaged in a glycosidic bond with the penultimate sugar, which is always galactose. A common structure found in this outer region is the trisaccharide NANA-α-(2→3)-Gal-β-(1→4)-GlcNAc (Fig.19).

Figure 19 Structure of the terminal nonreducing trisaccharide NANA-α-(2→3)-Gal-β-(1→4)-GlcNAc found in the outer region of several animal glycoproteins.

A common sequence also appears to exist in the core region of the carbohydrate side chains of glycoproteins of widely different origins (A4,A5). This conclusion is based on studies with asparagine oligosaccharides isolated from ovalbumin, α-amylase from *A. oryzae* and pineapple bromelain. Treatment of these glycopeptides with α-mannosidase to remove peripherally located mannose residues, yielded from all three glycoproteins the same compound, containing asparagine, N-acetylglucosamine and mannose in molar ratios 1:2:1. The residual mannose unit, which was not released by the α-mannosidase, was removed by a β-mannosidase. An exo-β-N-acetylglucosaminidase from Jack bean could then remove one of the hexosamine units. The β configuration of the glycosidic linkage between the two N-acetylglucosamine residues was confirmed by nuclear magnetic resonance measurements and the (1→4) position of the linkage was established by the Smith degradation method. For this purpose, the intact (enzymically undegraded) asparaginyl-oligosaccharides from the different glycoproteins were treated with $NaOH-NaBH_4$. As mentioned in a previous lecture (p.72), this resulted in complete cleavage of the glucosamine-asparagine linkage, with concomitant production of terminal glucosaminitol. Periodate oxidation, $NaBH_4$ reduction and hydrolysis in strong acid of the fragments thus produced, all yielded xylosaminitol.

```
     CH₂OH                              CH₂OH
      |                                  |
    HCNHAc                             HCNH₂
      |                                  |
     HOCH         (1) NaIO₄             HOCH
      |           (2) NaBH₄              |
     HCOR         (3) H⁺                HCOH
      |          ──────────→             |
     HCOH                               CH₂OH
      |
     CH₂OH
```

    4-*O*-substituted   xylosaminitol
    N-acetylglucosaminitol

Reference glucosaminitol compounds bearing substituents at positions 4 or 3 produced xylosaminitol and threitol, respectively. Final proof for the structure of the product isolated, after complete removal of mannose from the glycopeptides, was obtained when it was shown to be identical with the chemically synthesized GlcNAc-β-(1→4)-GlcNAc-β-(1→)-Asn (A6).

With the aid of the Smith degradation method and methylation studies, the penultimate glucosamine was found to be substituted by mannose at the 4 position. On the basis of these results it was proposed that the inner core region of the glycoproteins examined has the structure given in Fig.20.

The same sequence was found in the core region of ribonuclease B and of thyroglobulin (A7). An *N*-acetylgluco-

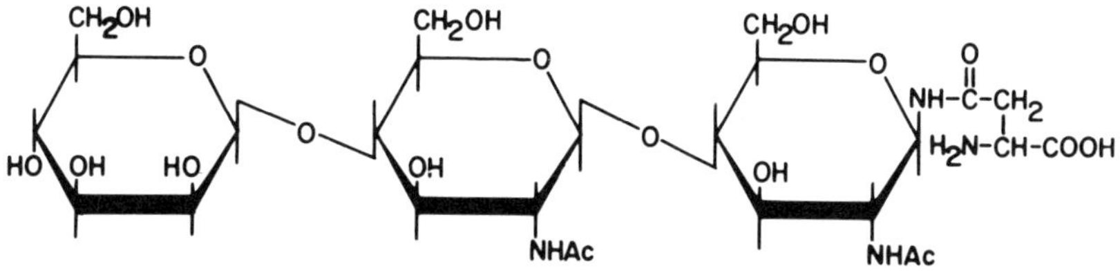

Figure 20 Structure of the inner core region of several glycoproteins

samine disaccharide joined to an amide nitrogen of asparagine probably also occurs as part of the complex heterosaccharides present in fetuin, immunoglobulin A and immunoglobulin G. This structure seems, therefore, to be a rather common feature of many glycoproteins.

In some glycoproteins, however, only one N-acetylglucosamine residue (and not two, as shown in Fig.20) is present in the linking region. This appears to be the case for the glycopeptides from soybean agglutinin and transferrin.

The examples I have given you serve to demonstrate that there are certain general rules for sugar assembly in glycoproteins. Therefore, the structural variation of their carbohydrate units is limited and is much smaller than could be expected on theoretical grounds (see Lecture 1, pp.7-9).

THE ASN-X-SER(THR) SEQUENCE

Examination of a large number of glycopeptides with the GlcNAc-Asn linking group where additional amino acids were present, revealed a common sequence next to the asparagine. This sequence is either Asn-X-Ser or Asn-X-Thr, with the carbohydrate linked to the asparagine (Table 9). It should be emphasized, however, that not all such sequences in proteins carry a carbohydrate side chain on the asparagine residue. In other words, a protein which possesses the requisite structural features for glycosylation may nevertheless not undergo this type of reaction. In fact, the sequence Asn-X-Thr (or Ser) occurs commonly also in proteins which lack sugar. Thus, bovine ribonuclease A and B possess an identical amino acid sequence which contains 11 Asn residues and the tripeptide, Asn(34)-Leu-Thr, but only in ribonuclease B is a carbohydrate chain attached to Asn 34.

R. D. Marshall and A. Neuberger suggested in 1968 that in the sequence Asn-X-Ser (or Thr), a hydrogen bond is formed between the carbonyl of the side chain of asparagine and the hydroxyl group of the hydroxyamino acid (Fig.21). This type of hydrogen bonding might be expected to reduce the acid dissociation constant of the amide group of the asparagine residues and thereby facilitate replacement of

Table 9

Amino acid sequences around asparagine-linked carbohydrate chains of several glycoproteins[a]

Simple carbohydrate chains	
Avidin (hen)	Leu-Gly-Ser-<u>Asn</u>-Met-Thr-Ile
IgM immunoglobulin, human	Leu-Tyr-<u>Asn</u>-Val-Ser-Leu
Ovalbumin	Glu-Lys-Tyr-<u>Asn</u>-Leu-Thr-Ser
Ribonuclease B (bovine)	Lys-Ser-Arg-<u>Asn</u>-Leu-Thr-Lys
Ribonuclease (porcine)	Ser-Arg-Arg-<u>Asn</u>-Met-Thr-Gln
Thyroglobulin (human)	Ala-Leu-Glu-<u>Asn</u>-Ala-Thr-Arg
Complex carbohydrate chains	
α_1-Acid glycoprotein	Pro-Ile-Thr-<u>Asn</u>-Ala-Thr-Leu
	Glu-Glu-Tyr-<u>Asn</u>-Lys-Ser-Val
	Phe-Thr-Pro-<u>Asn</u>-Lys-Thr-Glu
	Cys-Ile-Tyr-<u>Asn</u>-Thr-Thr-Tyr
	Gln-Arg-Glu-<u>Asn</u>-Gly-Thr-Ile
IgM immunoglobulin, human	Phe-Gln-Glx-<u>Asn</u>-Ala-Ser-Ser
Ribonuclease (porcine)	Ser-Ser-Ser-<u>Asn</u>-Ser-Ser-Asn

[a] Modified from R. G. Spiro (A2). The Asn residue to which the carbohydrate is attached is underlined. Simple carbohydrate units consist only of mannose and N-acetylglucosamine residues, while complex units contain in addition galactose, sialic acid, and sometimes also fucose and N-acetylgalactosamine.

one of the amide hydrogens by sugar (3).

Because an asparagine residue destined to undergo glycosylation occurs in only a relatively limited number of amino acid sequences, e.g. one out of 11 in ribonuclease B,

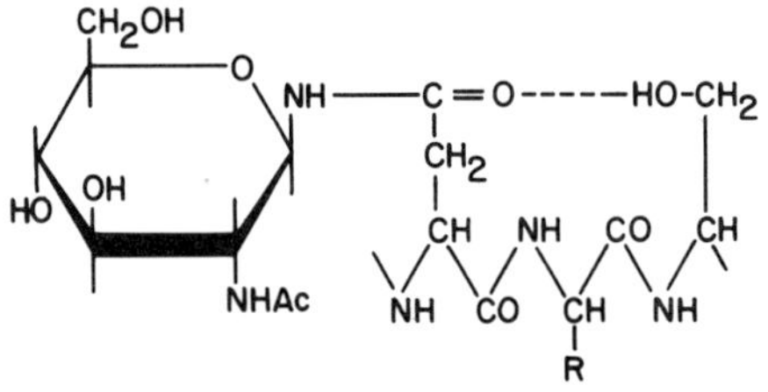

Figure 21   Schematic model of the Asn-X-Ser sequence found in glycoproteins containing the GlcNAc-Asn linkage. Note the hydrogen bond between the hydroxyl of serine and the β-carbonyl of asparagine. For a figure of the space filling model of the same sequence, see p.678 in Ref.3.

formation of linkages of this type is clearly under direct, as well as indirect, control of the genome. This circumstance is not unique: conversion of proline to hydroxyproline by collagen proline hydroxylase requires that the amino acids occur in the sequence .... Y-Pro-Gly...., where Y may be any of a number of amino acids apart from glycine. Similarly, the ability of hydroxymethylcytosine in the DNA of *E. coli* T 2 bacteriophage to accept glucosyl groups (see p.20) is markedly affected by the nature of the neighbouring nucleotides.

As we have just seen, glycosylation of the asparagine requires the presence of an hydroxyamino acid, serine or threonine, which in the biosynthesis of the polypeptide chain is inserted two polymerization steps later than the asparagine.

Attachment of the N-acetylglucosamine residue to asparagine cannot, therefore, proceed simultaneously with the insertion of this amino acid into the growing polypeptide chain.

MUCINS

In mucins, such as ovine or bovine submaxillary glycoproteins (OSM or BSM, respectively), the following unit occurs several hundred times in each molecule:

NANA-α-(2→6)-GalNAc-α-(1→)-Ser (or Thr)

Treatment of purified OSM under mild conditions (e.g. 1 hr at 80°C with 0.01 M mineral acid), liberates sialic acid quantitatively. The sialic acid accounts for 25% of the weight of the glycoprotein.

The sialic acid can also be enzymically cleaved from the glycoprotein by action of neuraminidase from *Vibrio cholerae* or *Clostridium perfringens*. The principal structure of the macromolecule appears to be unaltered by the removal of the sialic acid residues, the new terminal units being N-acetylgalactosamine. These in turn are susceptible to the action of α-N-acetyl hexosaminidase.

The intact disaccharide, NANA-α-(2→6)-GalNAc has been obtained in 40% yield by brief treatment of unmodified OSM with hot barium hydroxide. Under these conditions, the disaccharide is released by the β elimination reaction, which we discussed at length earlier (pp.74-78).

ANTIFREEZE GLYCOPROTEINS

A disaccharide side chain somewhat similar to that present in the submaxillary mucins is found in the antifreeze glycoproteins. Its structure is Gal-β-(1→3)-GalNAc-α-(1→)-Thr (Fig.4, p.39). The assignment is based, among others, on the use of galactose oxidase, which oxidizes the C-6 hydroxyl in galactose and N-acetylgalactosamine not only in their free form, but also when the monosaccharides are glycosidically linked (Fig.22).

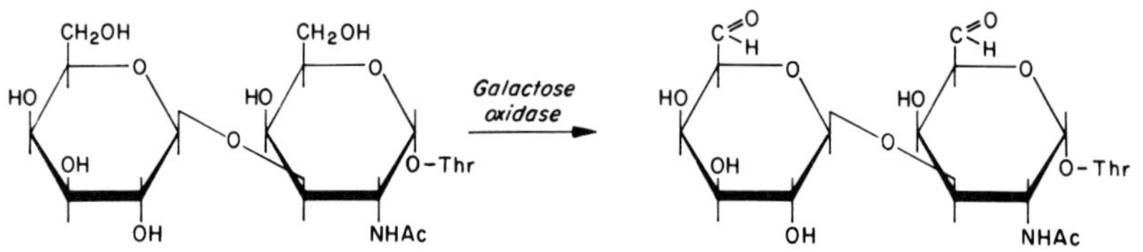

Figure 22 Oxidation of the primary hydroxyl groups in the disaccharide units of antifreeze glycoprotein by galactose oxidase (A8).

Interestingly, this enzymic oxidation did not impair the antifreeze properties of the glycoprotein. The activity was lost by the conversion of the newly formed

C-6 aldehydes to negatively charged groups either by oxidation to carboxyl groups with halogen or by formation of the bisulfite addition products (A8).

COLLAGEN AND BASEMENT MEMBRANE

Collagen and the basement membranes contain a disaccaride (Fig.23) O-glycosidically linked to the δ-hydroxyl group of several hydroxylysine residues in the polypeptide chain.

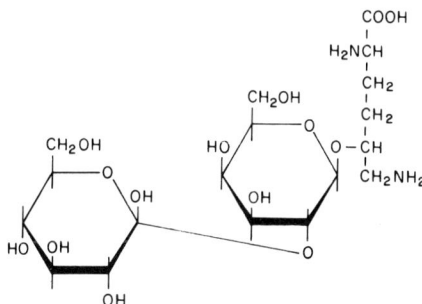

Figure 23   Structure of the hydroxylysine-linked disaccharide unit (Glc-α-(1$\rightarrow$2)-Gal) of the collagens and basement membranes.

Studies of the chemical structure of the glomerular basement membrane are of particular interest, since pathological changes have been observed in the renal glomerulus in diseases such as diabetes (4). The membrane is made of glycoprotein material with a composition that indicates that it belongs to the collagen family. Thus, it is rich in glycine, hydroxyproline and hydroxylysine. However, important compositional differences between basement membranes and fibrillar collagens have been noted, particularly in their carbohydrate, half cystine and hydroxylysine content. Approximately 10% of the membrane consists of sugar residues, as compared to less than 1% in most vertebrate fibrillar collagens. Although glucose and galactose are the major saccharide components in both membrane and collagen, only the basement membrane contains appreciable amounts of mannose, hexosamines, sialic acid and fucose.

The basement membrane is readily digested and almost completely solubilized by treatment with bacterial collagenase to yield its carbohydrate in the form of two distinct types of glycopeptides. These could be further degraded with pronase to give low molecular weight peptides in which the disaccharide Glc-α-(1$\rightarrow$2)-Gal is attached by a β-glycosidic linkage to the hydroxyl group of hydroxylysine and larger glycopeptides in which branched heteropolysaccharide units made up of sialic acid, fucose, galactose, glucosamine and mannose are linked to asparagine residues. The carbohydrate of the glomerular basement membrane is about equally distributed by weight between these two types of units, and there are 10 disaccharides for every heteropolysaccharide.

The hydroxylysine-linked carbohydrate unit has been found in other basement membranes and in a large number of collagens from vertebrate and invertebrate sources. In basement membranes such units are present in larger numbers than in fibrillar collagens. The density of the disaccharide units on the peptide chain appears to be inversely related to the morphologic organization of the collagen as seen under the electron microscope. The numerous, bulky polysaccharide units of the basement membrane may interfere with the packing of the peptide chains necessary for fibril formation. In fibrillar collagens, "hole" regions apparently exist to accommadate the smaller number of carbohydrate units.

Basement membranes isolated from human diabetic glomeruli were found to have a composition distinctly different from that of basement membranes from healthy subjects. In particular the diabetic membranes showed a marked increase in the number of glucosyl-galactose disaccharide units (4).

REFERENCES

Reviews

1. Microheterogeneity and function of glycoproteins, L. W. Cunningham *in* Glycoproteins of Blood Cells and Plasma (Eds. G. A. Jamieson and T. J. Greenwalt), J. B. Lippincott Co., Philadelphia and Toronto, 1971, pp.16-34.

2. Heterogeneity of the carbohydrate groups of glycoproteins,
R. Montgomery *in* Glycoproteins (Ed. A. Gottschalk), 2nd ed., part A, Elsevier, 1972, pp.518-528.

3. Glycoproteins,
R. D. Marshall, Ann. Rev. Biochem. 41, 673-702 (1972).

4. Biochemistry of the renal glomerular basement membrane and its alterations in diabetes mellitus,
R. G. Spiro, New Engl.J.Med. 288, 1337-1342 (1973).

Specific articles

A1. Microheterogeneity and paucidispersity of glycoproteins. Part I. The carbohydrate of chicken ovalbumin,
C. C. Huang, H. E. Mayer, Jr. and R. Montgomery, Carbohyd. Res. 13, 127-137 (1970).

A2. Glycoproteins,
R. G. Spiro, Adv. Prot. Chem. 27, 399, 368-370 (1973).

A3. Structure of the O-glycosidically linked carbohydrate units of fetuin,
R. G. Spiro and V. D. Bhoyroo, J. Biol. Chem. 249, 5704-5717 (1974).

A4. Pineapple α- and β-D-mannopyranosidases and their action on core glycopeptides,
Y. T. Li and Y. C. Lee, J. Biol. Chem. 247, 3677-3683 (1972).

A5. Common structural unit in asparagine-oligosaccharides of several glycoproteins from different sources,
Y. C. Lee and J. R. Scocca, J. Biol. Chem. 247, 5753-5758 (1972).

A6. The synthesis of a di-N-acetylchitobiose asparagine derivative, 2-acetamido-4-O-(2-acetamido-2-deoxy-β-D-glucopyranosyl)-1-N-(4-L-aspartyl)-2-deoxy-β-D-glucopyranosylamine,
M. Spinola and R. W. Jeanloz, J. Biol. Chem. 245, 4158-4162 (1970).

A7. A β-mannosidic linkage in the unit A oligosaccharide of bovine thyroglobulin,
A. L. Tarentino, T. H. Plummer, Jr., and F. Maley,
J. Biol. Cehm. 248, 5547-5548 (1973).

A8. Structure and role of carbohydrate in freezing point-depressing glycoprotein from antarctic fish,
J. R. Vandenheede, A. I. Ahmed, and R. E. Feeney,
J. Biol. Chem. 247, 7885-7889 (1972).

7

GLYCOPROTEIN BIOSYNTHESIS - I

The biosynthesis of conjugated macromolecules such as glycoproteins raises some problems which are not encountered in studies of simple proteins or simple polysaccharides. One of the main problems concerns the order in which the protein and the carbohydrate moieties are synthesized. In common with studies of other biological substances, there is also the question of cellular location and molecular mechanisms involved. Considerable information is accumulating about the intracellular sites of glycoprotein biosynthesis and about the sequence of events in the enzymic assembly processes (1,2). Little is known, however, about the tertiary structure, folding and conformation of glycoproteins in different environments.

STUDIES WITH INTACT CELLS

We shall first consider where the biosynthesis of glycoproteins occurs in the cell. Much of our knowledge on this question is based on studies with intact animals or on data obtained with organ perfusion techniques or tissue slices. Such studies usually involve exposure of the tissue to a labelled monosaccharide precursor (most commonly glucosamine, galactose, mannose or $\underline{L}$-fucose) in the absence or presence of inhibitors of protein synthesis, followed by examination of incorporation of the label into protein-bound carbohydrate. To obtain information on the relationship between the synthesis of the carbohydrate side chains and that of the polypeptide backbone, parallel experiments with labelled amino acids - most commonly with leucine - have been carried out. Incorporation of the label has been followed by two methods: (a) electron microscope radioautography, and (b) measurement of protein-bound radioactivity in subcellular fractions isolated by centrifugation of tissue homogenates.

The results obtained indicate that (a) the peptide backbone of glycoproteins is assembled on membrane bound ribosomes and (b) that most of the carbohydrate is incorporated into glycoproteins following release of the peptide backbone from the ribosomes. There is thus a clear separation in time and space between the biosynthesis of the polypeptide backbone of glycoproteins and the biosynthesis of their carbohydrate side chains.

A central role in the biosynthesis of the carbohydrate side chains of glycoproteins is performed by a subcellular organelle, the Golgi apparatus, or Golgi body (3,4). This structure consists of a group of membranous sacs, or a complex of interconnecting membranes. It was originally observed in 1898 by the Italian cytologist C. Golgi, who, by means of a special silver stain he had developed, showed that there is a reticular structure in the cytoplasm of certain nerve cells. The Golgi apparatus is now known to be present in all animal cells, and to serve as the primary site for the synthesis of large carbohydrates and for the packaging of intracellularly produced macromolecules. It thus functions in the biosynthesis of soluble glycoproteins, as well as in the formation of membranes, including those of storage and secretory vesicles.

The time course of incorporation of labelled monosaccharide precursors into the glycoproteins of subcellular organelles follows three general patterns (Table 10).
(a) Incorporation of radioactive mannose into protein appears

to parallel the pattern of incorporation of radioactive amino acids: there is rapid incorporation into rough-surfaced endoplasmic reticulum, followed by transfer to smooth surfaced endoplasmic reticulum and Golgi apparatus and final appearance of the label in terminal-pool protein (Pattern A). Both amino acid and mannose incorporation are sensitive to inhibition by puromycin. These findings reflect the fact that mannose occurs in the oligosaccharide core near the GlcNAc-Asn linkage region (Fig.20, p.107) and incorporation of this monosaccharide therefore takes place early in the biosynthetic process, shortly after the peptide itself has been assembled on the ribosomes. Puromycin inhibits peptide formation and there is subsequently no substrate to act as acceptor for mannose.

(b) Radioactive sialic acid, L-fucose and galactose are all incorporated primarily in the smooth-surfaced endoplasmic reticulum and Golgi apparatus and are then transferred to terminal-pool protein (Pattern B). This is to be expected for sugars that only occur at or near the non-reducing termini of oligosaccharide prosthetic groups and are therefore incorporated into glycoprotein in the final stages of the biosynthetic process. The incorporation of L-fucose and galactose is not appreciably inhibited by puromycin, since there is an adequate reserve of unfinished glycoprotein within the endoplasmic reticulum to act as acceptor for these sugars. This reserve is maintained for an appreciable length of time after protein synthesis has been inhibited by puromycin. Radioactive $\underline{\text{L}}$-fucose is an excellent label for the Golgi apparatus because it is not metabolized to any large extent to amino acids or to other sugars.

(c) Radioactive glucosamine is incorporated simultaneously into both rough- and smooth-surfaced endoplasmic reticulum and is then transferred to terminal-pool protein (Pattern C). This is consistent with the fact that N-acetylglucosamine appears in three locations in the oligosaccharide prosthetic group (Fig. 18, p.104), namely the GlcNAc- Asn linkage region, the core and the terminal non-reducing trisaccharides. Thus glucosamine incorporation occurs throughout the biosynthetic process.

There is considerable controversy as to whether the N-glycosidic linkage between N-acetylglucosamine and asparagine in glycoproteins is formed before the completion of the peptide and its release from the ribosome, or subsequent to this completion and release. Evidence from work with liver and plasmocytoma indicates that a small amount of N-acetylglucosamine becomes incorporated into peptide which is

Table 10

Incorporation of label into organelle glycoproteins

Pattern type	Radioactive precursor	Time course of incorporation into glycoprotein	Interpretation
A	Leucine, Mannose	Golgi, SG,EC / RER (curves)	RER → Golgi → SG, EC or PM ↑ label

B	Sialic acid, L-Fucose, Galactose	SG, EC or PM / Golgi, not RER (curves)	RER → Golgi → SG, EC or PM
			        ↑
			        label

C	Glucosamine	SG, EC or PM / RER and Golgi (curves)	RER → Golgi → SG, EC or PM ↑ ↑ label label
Hours after pulse		0 1 2 3	

Abbreviations: RER: rough-surfaced endoplasmic reticulum; SG, secretory granules; EC, extracellular space; PM, plasma membrane.

The above time courses are idealized and only qualitatively accurate; there is appreciable quantitative variation from one tissue or species to the other. From H. Schachter, Symposium on Glycoconjugates, Lille 1973.

still nascent on ribosomes. There is the remote possibility that GlcNAc-Asn is incorporated into protein as such, but all quests for an GlcNAc-Asn "activating enzyme" have ended with failure. Recently, R. D. Marshall has studied a liver enzyme that transfers N-acetylglucosamine from UDP-GlcNAc to Asn 34 in ribonuclease A, converting it into ribonuclease B. With respect to the formation of other types of carbohydrate peptide linkages (e.g. GalNAc-Ser, Gal-Hyl or Xyl-Ser) the situation is much clearer. This I shall discuss somewhat later.

The kinetic data summarized in Table 10 indicate the stepwise incorporation of monosaccharides into oligosaccharide prosthetic groups while nascent peptides move from rough-surfaced endoplasmic reticulum to smooth-surfaced endoplasmic reticulum and Golgi apparatus and then finally to the terminal pool. The nature of this terminal pool depends on the tissue under study; thus in the thyroid the terminal pool represents thyroglobulin stored in the colloid. In secretory cells which do not store their secretory products, the terminal pool is the secreted product, e.g. in liver it is represented by plasma glycoproteins. In all cells, presumably, some of the terminal pool must be membrane-bound glycoprotein; this is especially apparent in non-secretory cells such as HeLa cells and duodenal

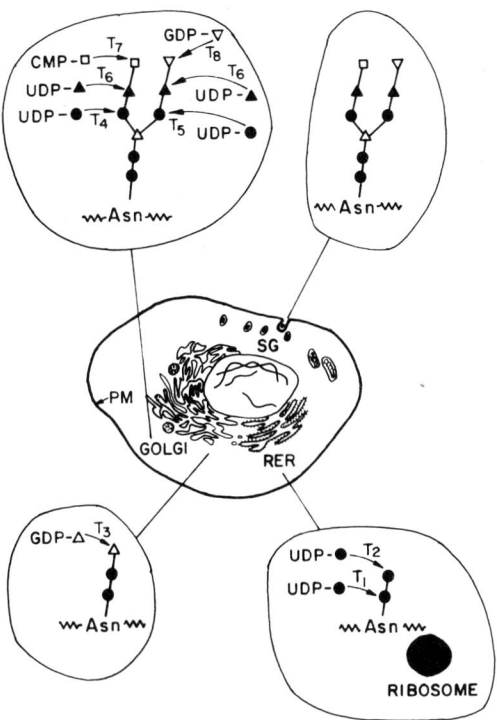

Figure 24 Diagram illustrating the movement of glycoprotein molecules through a cell such as of liver. The polypeptide core of the glycoprotein is assembled on ribosomes in the rough endoplasmic reticulum (RER); some carbohydrate may be incorporated into peptide at this stage by the membrane-bound glycosyltransferases. The peptide is released from the ribosome into the intravesicular channels of the endoplasmic reticulum and traverses these channels from rough endoplasmic reticulum to smooth endoplasmic reticulum to Golgi apparatus. Sugars are incorporated sequentially throughout this traverse by a multi-glycosyltransferase system firmly attached to membrane (T_1, T_2, T_3, ... T_n). The terminal sugars of the side chain group (usually sialic acid, fucose and galactose) are incorporated into glycoprotein in or near the Golgi apparatus. The completed glycoprotein is packaged into secretory granules (SG) by the Golgi apparatus; these granules then break off the Golgi apparatus and migrate

toward the plasma membrane (PM). The membrane of
the secretory granules fuses with plasma membrane
and the fused portion then breaks down to release
glycoprotein into the extracellular space. The
fusion process is believed to be a mechanism of
generating new plasma membrane. Not shown in the
diagram is the role that the Golgi apparatus is
believed to play in generating intracellular
membrane for organelles such as lysosomes. The
glycoprotein components of such intracellular
membranes appear to be assembled in a manner analogous to the secreted glycoproteins. (●, N-acetylglucosamine; △, mannose; ▲, galactose; ▽, $\underline{L}$-fucose; ◻, sialic acid).

columnar cells, in which the terminal pool is the plasma membrane glycoprotein.

It is of interest to point out that labelled liver glycoproteins are readily released from both rough- and smooth-surfaced microsomes by ultrasonic oscillation, implying that these materials are within the cisternal spaces of the endoplasmic reticulum. However, some newly synthesized glycoprotein is also firmly attached to the microsomal membrane and it is possible that the difference in binding to the membrane may reflect the ultimate destination of the glycoprotein, either in the extracellular secretion or as part of the intracellular membrane system.

A diagramatic representation of our current views of the biosynthesis is given in figure 24, modified from Schachter and Roden (2).

BIOSYNTHESIS OF IMMUNOGLOBULINS

Studies on the biosynthesis of immunoglobulin light chains secreted by myeloma cell lines of mice (e.g. line MOPC-46) have provided further information about the mechanism of intracellular assembly of the carbohydrate units of glycoproteins (A1). The carbohydrate units of the completed light chains secreted from MOPC-46 plasmacytoma cells contain N-acetylglucosamine, mannose, galactose, $\underline{L}$-fucose and sialic acid in the molar ratio 3:4:4:2:2. The units appear to be similar to the oligosaccharide structures attached to human immunoglobulin heavy chains, although they are probably more highly branched. After subcellular fractionation of the tumor cell, light chains were released from the membranes of the rough endoplasmic reticulum by mild treatment with detergent. This material, purified by ion exchange chromatography, was found to contain 2-3 moles of N-acetylglucosamine, 3-4 moles of mannose and only traces of other sugars. The product isolated from smooth membranes contained, in addition, about half of the galactose content of the secreted glycoprotein, indicating that this sugar was attached subsequent to the completion of a core oligosaccharide containing only N-acetylglucosamine and mannose. There is, however, evidence that N-acetylglucosamine transfer occurs both at the rough and the smooth membranes (which contain Golgi membranes).

Radioautographic studies in these cell lines using labelled leucine, mannose or galactose, clearly showed that migration from the rough endoplasmic reticulum to the Golgi apparatus occurs after incorporation of the mannose into polypeptide but before attachment of galactose. The completion of the carbohydrate chains then proceeds in the membrane stacks of the Golgi apparatus.

SYNTHESIS OF SUGAR CONSTITUENTS

Like most monosaccharides found in nature, most of the sugar constituents of carbohydrates in animal cells, with the exception of sialic acid, are synthesized as nucleotide-linked derivatives from nucleotide-linked precursors (Fig.25), the best known of which is uridine diphosphate glucose (p.19). Relatively few, e.g. N-acetylglucosamine, may arise from products of intermediate metabolism.

The recognition of the central role played by sugar nucleotides in carbohydrate metabolism, which stemmed from the pioneering work of Leloir and his co-workers, was of prime importance in paving the way for our understanding of the biosynthesis of simple and complex polysaccharides - including glycoproteins (cf. pp.18-22). The nucleoside

GLYCOPROTEIN BIOSYNTHESIS I

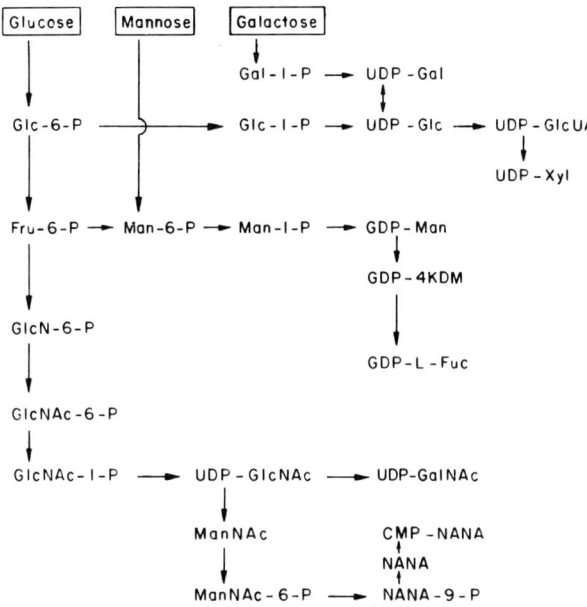

Figure 25 Pathways of formation of sugar nucleotides in animals.

diphospho moiety can be considered as a handle, which holds the sugar for transformation or transfer and appears to confer specificity on the enzymes catalyzing these reactions.

Nucleotide-linked sugars can undergo several types of modification reactions, or transformations, including epimerization, oxidation, decarboxylation, reduction and rearrangement (5). In all of these reactions nicotine amide adenine dinucleotide (NAD or DPN) or nicotine amide adenine dinucleotide phosphate (NADP or TPN) or both, are required.

In their other role, nucleotide-linked sugars serve as donors of sugar residues in the formation of oligo- and polysaccharides, lipopolysaccharides, glycoproteins and glycolipids (6). In the same organism, different nucleotide diphosphates may be used for activation of different sugars - e.g. glucose as UDP-Glc, and mannose as GDP-Man. Sometimes, however, the same sugar may be attached to different nucleotide diphosphates. Thus, in bacteria UDP-Glc may be utilized for the biosynthesis of lipopolysaccharides, whereas ADP-Glc is the precursor of glycogen; in contrast, in animals glycogen is formed from UDP-Glc.

The use of different nucleotides as carriers of monosaccharides may be advantageous to the organism in that it separates pathways of synthesis and offers a means for their independent control (5).

The most common precursor of the sugar moieties of complex carbohydrates is glucose. Formation of such complex compounds from glucose is the result of a number of reactions.

1. Synthesis of monomer units, either directly from glucose (e.g. mannose) or via modifications (reaction 3 below).
2. Activation of monomers (by conversion into nucleotide-linked sugars).
3. Modification of the nucleotide-linked sugar (e.g. UDP-Glc → UDP-Gal).

4. Transfer to a polymeric acceptor, either directly or via a lipid carrier.

As pointed out, all interconversions of glucose to the sugars found in glycoproteins occur at or prior to the nucleotide sugar state. Recent work has shown, however, that epimerization reactions are not exclusively limited to the precursor stage. Thus the L-guluronic acid component of alginic acid is formed by epimerization of D-mannuronic acid after incorporation of the latter into the polymer (A2), and similarly the L-iduronic acid residues of heparin are formed at the polymer level by epimerization of D-glucuronic acid residues (A3) (see also lecture 16).

BIOSYNTHESIS OF 6-DEOXY SUGARS (7)

Before discussing the results of studies of glycoprotein synthesis in cell free systems, let us see in more detail how some typical monosaccharide constituents of glycoproteins are formed.

L-Fucose is formed from mannose, via the nucleotide-linked sugar GDP-Man. This involves epimerization at C-3 and C-5 and reduction at C-6 of the mannose molecule.

```
        CHO                  CHO
         |                    |
        HOCH                 HOCH
         |                    |
        HOCH                 HCOH
         |                    |
        HCOH                 HCOH
         |                    |
        HCOH                 HOCH
         |                    |
        CH₂OH                CH₃

       Mannose             L-Fucose
                      (6-deoxy-L-galactose)
```

In the first step, GDP-Man is synthesized by the following reaction

$$\text{GTP + Man-1-P} \rightleftharpoons \text{GDP-Man + PP}_i$$

The reaction is reversible, with the equilibrium to the left, i.e. in the direction of pyrophosphorylysis of GDP-Man, but *in vivo* it proceeds to the right. This is because of the ubiquitous presence of pyrophosphatases, which hydrolyse the inorganic pyrophosphate (PP_i) formed, and shift the equilibrium to the synthesis of GDP-Man.

The transformation of GDP-Man to GDP-L-Fuc involves two stages. The first stage results in the formation of a nucleotide-linked 4-keto-6-deoxy intermediate; in this case GDP-4-keto-6-deoxy-D-mannose (GDP-4KDM). In the second step, this intermediate is converted into GDP-L-fucose (Fig.26).

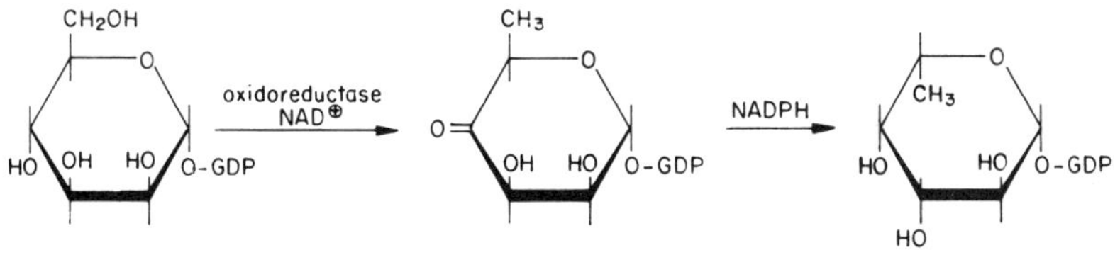

Figure 26 Conversion of GDP-Man into GDP-L-Fuc via the 4-keto-6-deoxy intermediate.

The 4-keto compound is a key intermediate in this type of reaction, common for the biosynthesis not only of deoxyhexoses but also of 3,6-dideoxyhexoses, 4-amino-4, 6-dideoxyhexoses and related compounds. The reaction is catalysed by enzymes referred to as oxidoreductases and is irreversible. Once a nucleotide-linked hexose becomes converted to the 4-keto derivative, the sugar is no longer available for the main metabolic pathways of energy production.

The conversion into the 6-deoxyhexose, in this case L-fucose, is the result of epimerization at C-3 and C-5 of the 4-keto-6-deoxy derivative, followed by stereospecific reduction at C-4.

The reaction mechanism of deoxy sugar biosynthesis has been elucidated with enzymes from *E. coli*, which convert TDP-D-Glc to the corresponding L-rhamnose (6-deoxy-L-mannose) derivative, TDP-L-Rha.

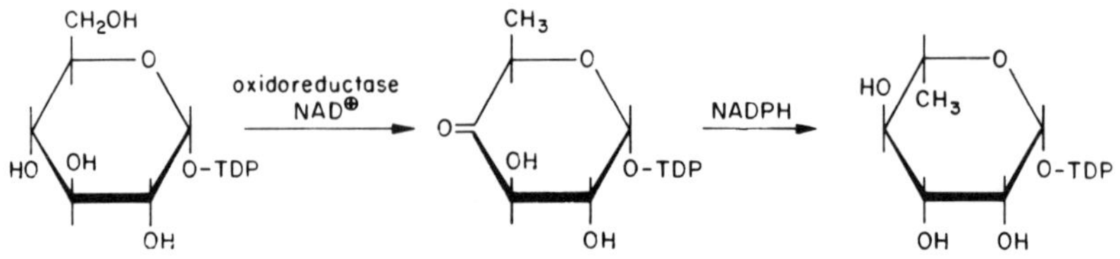

Figure 27 Conversion of TDP-Glc into TDP-L-Rha via the 4-keto-6-deoxy intermediate.

This sequence of reactions (Fig.27) is analogous to the transformation of GDP-Man to GDP-L-Fuc. The first step involves oxidation at C-4 and conversion of the primary alcoholic group at C-6 to a methyl group. The elucidation of the reaction mechanism was accomplished by using a variety of approaches which included preparation of selectively tritiated substrates to trace the fate of the tritium during the enzymic reaction. It was thus found that formation of the 4-keto-6-deoxy compound proceeds by an intramolecular hydrogen transfer from C-4 to C-6 (Fig.28).

This was established in a study with TDP-Glc labelled specifically at C-4 with tritium. The labelled sugar nucleotide was obtained from synthetic glucose-4-T, which was converted by hexokinase and ATP to glucose-6-phosphate-4-T. Treatment of the latter compound with phosphoglucomutase afforded glucose-1-phosphate-4-T, which reacted with deoxy thymidine triphosphate (TTP) in the presence of TDP-Glc pyrophosphorylase, to yield TDP-Glc-4-T. When the preparation of TDP-Glc-4-T was incubated with the TDP-Glc oxidoreductase, the substrate was converted quantitatively to TDP-4-keto-6-deoxyglucose-6-T: the tritium originally present on C-4 of the glucose was transferred to C-6 of the 4-keto derivative. No exchange of tritium with the medium occurred during the reaction. An identical intramolecular hydrogen transfer was observed when TDP-Glc specifically labelled with deuterium at the C-4 of the glucose (TDP-Glc-4-D) was used as the substrate. These experimental findings are consistent with formation of a 5,6-glucoseen derivative, resulting from loss of water from C-5 and C-6 of the original sugar (Fig. 28). This loss of water is followed by reduction of the double bond between C-5 and C-6.

Homogeneous preparations of the TDP-Glc oxidoreductase from *E. coli* contained one mole of NAD^+ per mole of enzyme. All available evidence shows that the reactions catalyzed by the enzyme occur with the NAD^+ firmly bound to its surface. TDP-Glc is initially attacked by enzyme-NAD^+ to yield TDP-4-ketoglucose, with concurrent formation of enzyme-NADH (Fig.28). The 4-keto derivative of glucose (a 4-hexulose) is converted by β-elimination of water between C-5 and C-6 to

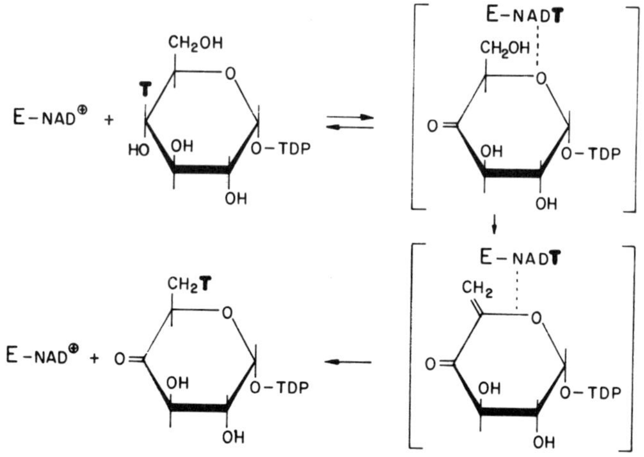

Figure 28 Proposed mechanism of oxidoreductase reaction. E stands for the enzyme. All intermediates given in parenthesis in this and other figures are hypothetical.

form the unsaturated glucoseen. This 5,6-glucoseen serves as hydrogen acceptor for the enzyme-NADH complex, with restoration at enzyme-NAD$^+$, and formation of the end product of the reaction, TDP-4KDG. The latter compound is more appropriately named TDP-6-deoxy-D-xylo-4-hexulose.

Chemical reduction of the 4-keto-6-deoxy compound affords a mixture of two epimeric 6-deoxy sugar derivatives, 6-deoxygalactose (D-fucose) and 6-deoxyglucose (quinovose), which differ in the configuration at C-4. However, in the enzyme catalyzed reaction only one product is obtained (L-rhamnose), which is different from those just mentioned.

GLYCOPROTEIN BIOSYNTHESIS I

After the formation of the 4-keto intermediate, it is still necessary to change the configuration at C-3 and C-5, and to reduce the keto group at C-4, in order to obtain the final product, TDP-L-Rha (Fig.29). At least two and possibly three additional enzymes participate in these reactions. As a first step, an enzyme (or enzymes) referred to as 3,5 isomerase, catalyzes epimerizations at carbons 3 and 5, probably via the enediol form. For example, in the first stage of the reaction, the keto-enol transformation results in double bond formation between C-3 and C-4 with loss of assymetry at C-3. Stereospecific rearrangement into the 4-keto compound, resulting in epimerization at C-3, leads to the same configuration at this carbon as in L-rhamnose. In a similar way, epimerization is assumed to proceed at C-5. It should be emphasized

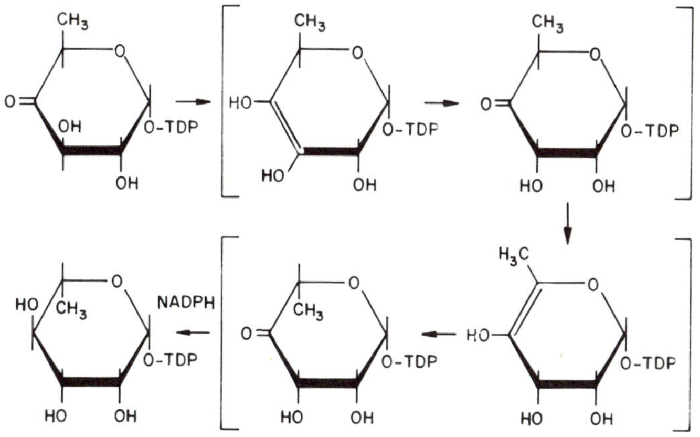

Figure 29 Proposed mechanism for the conversion of TDP-4-keto-6-deoxyglucose into TDP-L-rhamnose.

that these steps are hypothetical, as are all the intermediates in brackets in Fig.29. It is further assumed that the intermediates are bound to the enzyme.

The epimerizations are followed by a stereospecific reduction by an enzyme referred to as a reductase. In this reaction a stoichiometric amount of NADPH is required to reduce the keto group at carbon 4, resulting in release from the enzyme of TDP-L-Rha, the final product of the pathway.

Undoubtedly the same reaction mechanisms operate in the conversion of GDP-Man to GDP-L-fucose.

BIOSYNTHESIS OF GALACTOSE AND XYLOSE

There is now a considerable amount of evidence showing that 4-keto sugars also serve as intermediates in reactions not involved in 6-deoxyhexose biosynthesis. Among these reactions is the epimerization of glucose to galactose by UDP-Gal-4-epimerase, previously known as "galactowaldenase", and the UDP-glucuronic acid decarboxylase. The latter enzyme catalyzes the conversion of UDP-GlcUA to UDP-Xyl, which is the donor of xylose for the synthesis of the carbohydrate-peptide linkage in mucopolysaccharides.

As we have just hinted, galactose is formed via UDP-Glc. This is the result of an epimerization reaction which is NAD^+

linked. It occurs by an oxidation reduction reaction without any intermediates released from the surface of the enzyme. Thus, no exchange of tritium or oxygen-18 with the medium could be detected in studies of this reaction, and a 4-keto intermediate has not been isolated. There is, however, sufficient evidence for the participation of such an intermediate in the reaction (Fig.30).

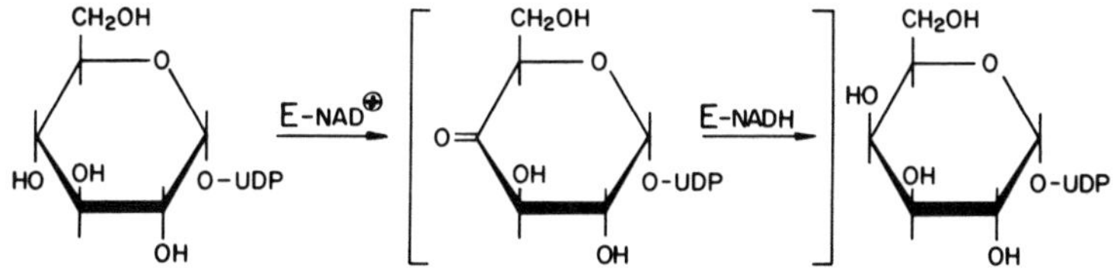

Figure 30 Mechanism of action of UDP-Gal-4-epimerase. E denotes the enzyme.

All that UDP-Gal-4-epimerase does is to flip over the 4-OH from one position to the other, converting an equatorial hydroxyl at C-4 in glucose to an axial one, as in galactose (Fig.31). The presence of the 4-axial hydroxyl introduces a certain degree of instability into the molecule of galactose as compared to glucose. This is probably the reason why the ratio of UDP-Glc to UDP-Gal at equilibrium in the enzyme catalyzed epimerization reaction is 3:1.

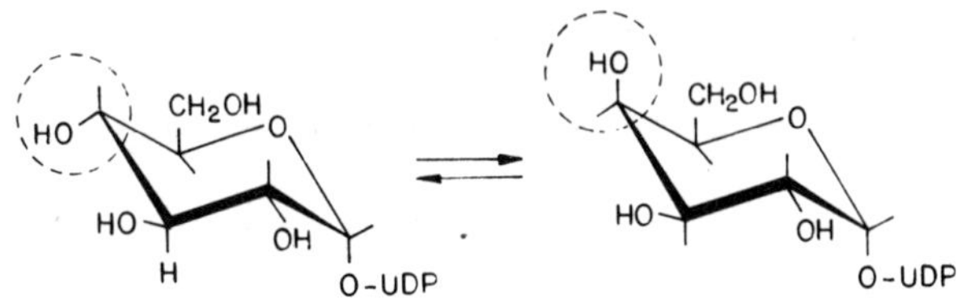

Figure 31 Conversion of UDP-Glc to UDP-Gal catalyzed by UDP-Gal-4-epimerase.

Other 4-epimerases are known, such as those that convert UDP-GlcNAc to UDP-GalNAc, UDP-Xyl to UDP-L-Ara or TDP-GlcNAc to TDP-GalNAc. Apparently, they operate by the same mechanism as UDP-Gal-4-epimerase.

REFERENCES

Reviews

1. Biosynthesis of glycoproteins and its relationship to heterogeneity,
    A. Gottschalk, Nature 222, 452-454 (1969).
    An excellent introduction to the subject.

2. The biosynthesis of animal glycoproteins,
    H. Schachter and L. Rodén *in* Metabolic Conjugation and Metabolic Hydrolysis (Ed. W. H. Fishman), Vol.3, Academic Press, 1973, pp.1-149.
    A comprehensive and most useful review of the field. Highly recommended.

3. The Golgi apparatus,
    M. Neutra and C. P. Leblond, Scien. Amer. 220 (2), 100-107 (1969).

4. Golgi apparatus: influence on cell surfaces,
 W. G. Whaley, M. Dauwalder and J. E. Kephart, Science
 175, 596-599 (1972).

5. Sugar nucleotides and the synthesis of carbohydrates,
 V. Ginsburg, Advanc. Enzymology 26, 35-88 (1964).
 *Although written a dozen years ago, it is still
 a most basic review of the subject with many
 stimulating ideas.*

6. Biosynthesis of saccharides from glycopyranosyl esters
 of nucleoside pyrophosphates ("sugar nucleotides"),
 H. Nikaido and W. Z. Hassid, Adv. Carb. Chem. Biochem.
 26, 351-483 (1971).
 *A thorough and comprehensive review, which deals with
 the biosynthesis of glycoproteins and other complex
 carbohydrates.*

7. Biological mechanisms involved in the formation of
 deoxy sugars: enzymatic hydrogen mediation,
 O. Gabriel, Adv. Chem. Series 117 (ACS), 387-410 (1973).

Specific articles

A1. Biosynthesis of the carbohydrate portion of immunoglobulin. Radiochemical and chemical analysis of carbohydrate moieties of two myeloma proteins purified from different subcellular fractions of plasma cells,
 F. Melchers, Biochemistry 10, 653-659 (1971).

A2. Biosynthesis of alginate. Part II. Polymannuronic acid
 C-5 epimerase from *Azotobacter vinelandii* (Lipman),
 A. Haug and B. Larsen, Carbohyd. Res. 17, 297-308
 (1971).

A3. Biosynthesis of L-iduronic acid in heparin: epimerization of D-glucuronic acid on the polymer level,
 U. Lindahl, G. Bäckström, A. Malmström and L. A. Fransson, Biochem. Biophys. Res. Commun. 46, 985-991 (1972).

8

THE SIALIC ACIDS

The sialic acids (1,2) are important constituents of glycoproteins as well as of gangliosides and milk oligosaccharides (Table 11), which deserve special consideration. Before describing their biosynthesis, let me say a few words about their chemistry.

In contrast to all the other sugars we have discussed, which are built of six or five carbons in a chain, the sialic acids are nine-carbon sugars. They are predominantly N- and O-acyl derivatives of the α-ketopolyhydroxyamino acid, known as neuraminic acid, which can be viewed as a condensation product of mannosamine and pyruvic acid. It was first isolated in its diacetyl form from bovine submaxillary mucin by G. Blix in Uppsala in 1936 and subsequently as neuraminic acid β-methyl glycoside from brain gangliosides by E. Klenk

SIALIC ACIDS

Table 11

Sialic acid-containing polymers

	Carbohydrate constituents						
	Sialic acid	Gal	Glc	Man	Glc-NAc	Gal-NAc	Fucose
Colominic acid (*E.coli*) (polyneuraminic acid)	+						
Milk oligosaccharides	+	+	+		+		+
Blood glycoproteins	+	+		+	+	+	+
Submaxillary mucins	+	+			+	+	+
Gangliosides	+	+	+			+	

in Köln in 1941. The most widely distributed sialic acid is N-acetylneuraminic acid named systematically 5-acetamido-3,5-dideoxy-$\underline{D}$-glycero-$\underline{D}$-galactononulosaminic acid (abbreviated as NANA or as NeuAc) (Fig.32). The sialic acids are readily

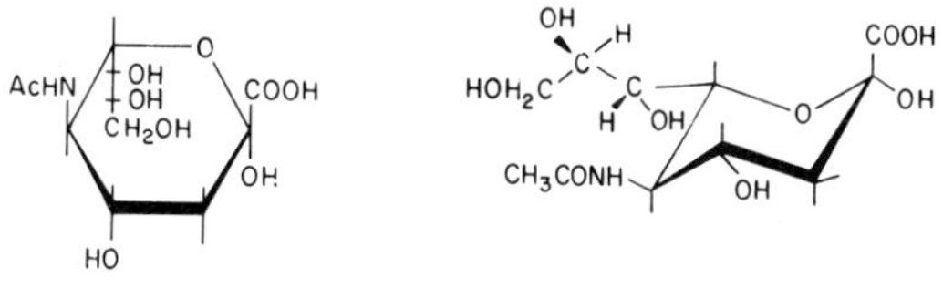

Figure 32 Structure of N-acetylneuraminic acid (NANA or NeuAc).

degraded by both acids and bases. The parent compound of this family, neuraminic acid, is unstable and has never been encountered in nature. To date 15 different sialic acids have been identified in nature (2,A1). However, the biological significance of the variations in their structure is not known. Some examples of naturally occurring sialic acids are given below. Straight-chain structure of neuraminic acid is also included, for comparison.

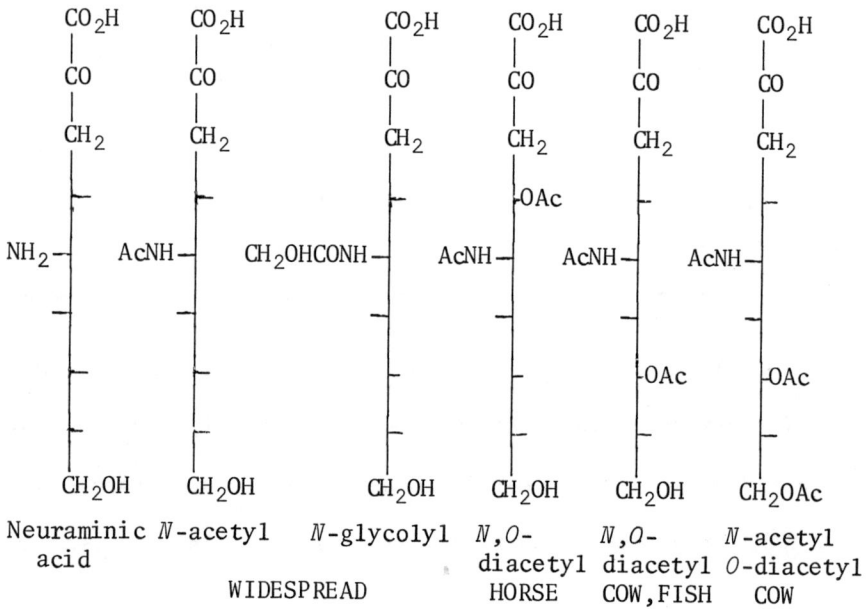

N-Acetylneuraminic acid and, most probably, all other sialic acids, occur in the pyranose form and have the $1C$ conformation (Fig.32).

The stereochemistry of the ketosidic bond of sialic acid has been elucidated by a study of the two anomeric methyl ketosides (Fig.33, I and II (A2)). In this study, the ketosides were converted by periodate-borohydride treatment

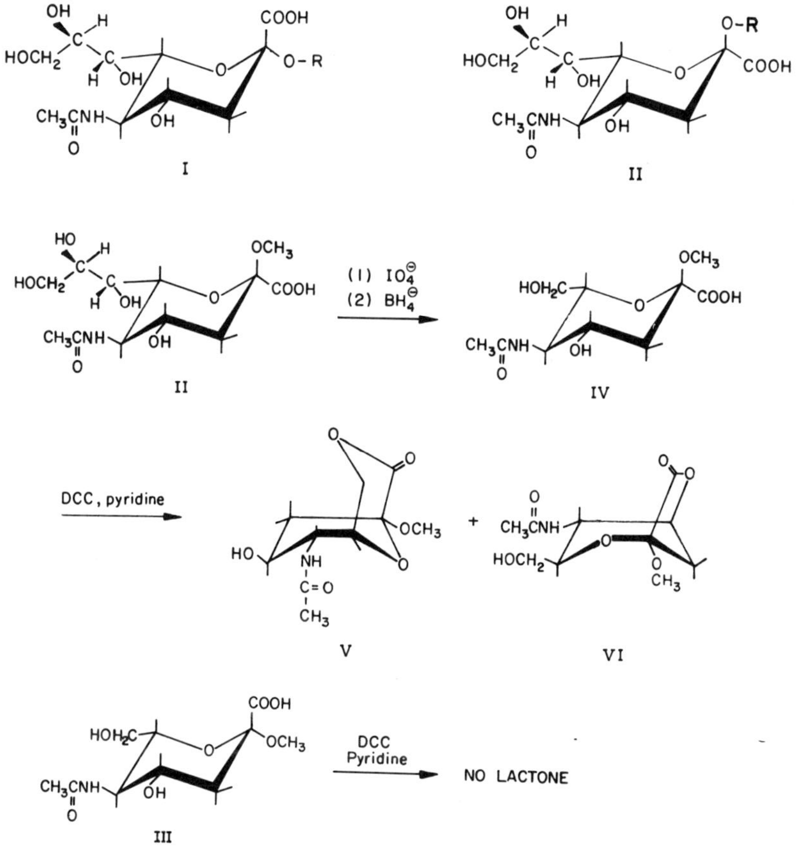

Figure 33 Reactions used in demonstrating the stereochemistry of the anomeric N-acetylneuraminic acid ketosides (from A2). I and II are the α and β-methyl ketosides of NANA, respectively. DCC denotes dicyclohexylcarbodimide.

to compounds III and IV, respectively, followed by lactonization of one of the isomers only. Indeed, examination of Dreiding models of compounds III and IV demonstrates that only one anomer is potentially able to lactonize, and this was accordingly assigned structure IV. This anomer and its precursor (II) were resistant to neuraminidase, whereas compounds I and III were hydrolyzed by the enzyme.

The configuration of the ketosidic bond of sialic acid in naturally occurring substances is thus of the less stable anomer I (Fig.33), with the ketosidic bond equatorial and the carboxyl group axial to the pyranoid ring. It is assigned the α-$\underline{D}$ configuration in accordance with the rules of nomenclature. However, in cytidine 5'-monophospho-N-acetylneuraminic acid, the substrate for sialyltransferase, the linkage is apparently β-ketosidic (Fig.37, p.151).

N-Acetylneuraminic acid can be synthesized either chemically, by condensation of 4,6-benzylidene-N-acylglucosamine and di-*tert*-butyl oxalacetate (in this reaction, epimerization at C-2 of glucosamine to mannosamine occurs) or by enzymic condensation of pyruvate and N-acetylmannosamine. Using suitable hexosamine derivatives as starting materials, N-glycolylneuraminic acid and a number of sialic acids not occurring in nature have been synthesized.

As already mentioned, acylneuraminic acids generally occupy the non-reducing ends of hetero-oligosaccharide chains in glycoproteins and glycolipids. In these molecules they are bound by α-glycosidic linkages to galactose, N-acetylgalactosamine or less often to a second N-acetylneuraminic acid molecule. An unusual compound is colominic acid produced by *E. coli*, shown to be a polymer of N-acetylneuraminic acid. Most sialic acids are readily released from their glycosidic linkages by neuraminidases or dilute acids (p. 58).

The nine-carbon backbone of acylneuraminic acids is formed *in vivo* by condensation of N-acetylmannosamine (or its 6-phosphate) with phosphoenolpyruvate. This reaction is catalyzed by N-acetylneuraminate (-9-phosphate) synthase. N-Acetylmannosamine may be formed by direct epimerization of N-acetylglucosamine by an enzyme which is found in many animal tissues. Usually, however, it is formed from UDP-GlcNAc, by a 2-epimerase, according to the following reaction:

$$\text{UDP-}N\text{-acetylglucosamine} \longrightarrow N\text{-acetylmannosamine} + \text{UDP}.$$

This epimerization reaction is clearly unique among the hexose inter-conversions involving nucleotide-linked sugars, since the monosaccharide product is free, and not nucleotide-bound. The mechanism of this reaction was investigated in detail by W. Salo and H. G. Fletcher (A3) who concluded, among other things, that UDP-N-acetylmannosamine is not an intermediate in the 2-epimerization reaction.

There is now evidence (A4) that this 2-epimerase reaction may proceed by a *trans* elimination of UDP, with formation of an intermediary unsaturated compound, a 2-acetamidoglucal, as shown in Fig.34.

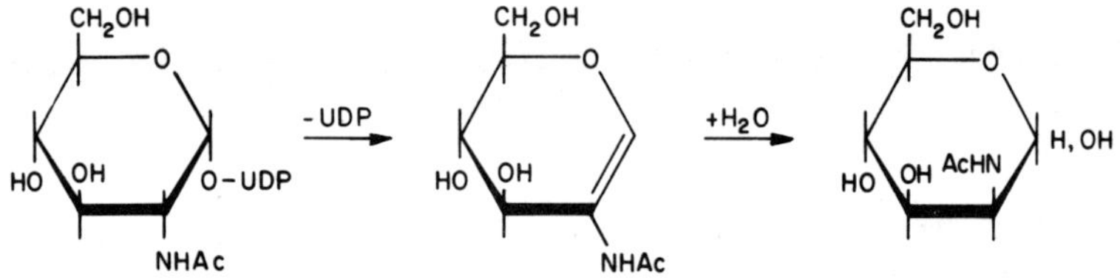

Figure 34 Proposed mechanism of action of UDP-N-acetyl-glucosamine-2-epimerase. I, UDP-GlcNAc; II, 2-acetamidoglucal; III, N-acetylmannosamine (A4).

The transformation of N-acetylmannosamine to N-acetylneuraminic acid in mammalian tissues, starts with the formation of N-acetylmannosamine-6-phosphate:

$$\text{ManNAc} + \text{ATP} \longrightarrow \text{ManNAc-6-P} + \text{ADP}$$

ManNAc-6-P then condenses with phosphoenolpyruvate (PEP), and the resultant NANA-9-phosphate is dephosphorylated to NANA (Fig.35).

N-Acetylneuraminic acid is produced both by vertebrates and invertebrates and by some bacteria. It is a precursor of N-glycolylneuraminic acid, of N-acetyl-4-O-acetylneuraminic acid and of N-acetyl-7- or 9-O-acetylneuraminic acids found in a variety of tissues. Formation of N-glycolylneuraminic

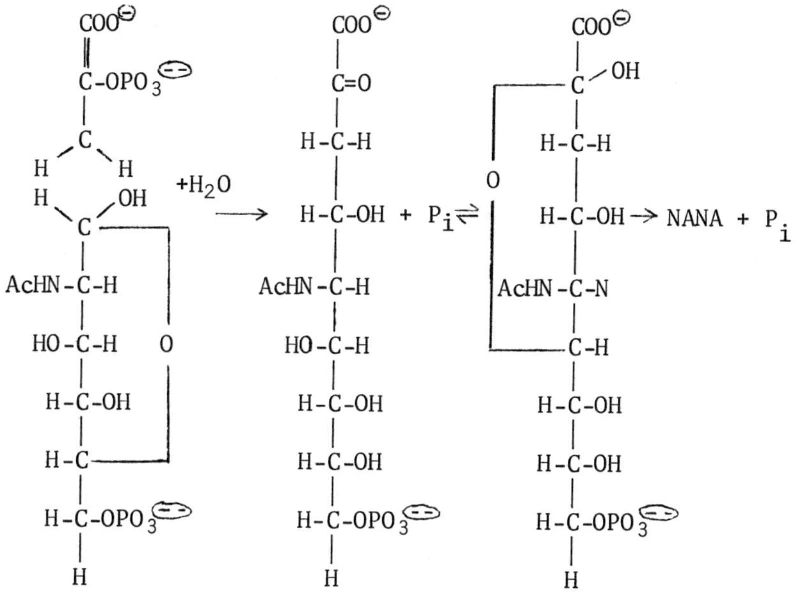

SIALIC ACIDS

Figure 35 Synthesis of NANA from N-acetylmannosamine-6-phosphate and phosphoenolpyruvate.

acid from NANA is by direct enzymic hydroxylation. In addition to oxygen, this reaction requires ascorbate or NADPH and Fe^{++}. The acetyl donor in this reaction is acetyl coenzyme A. The enzyme acetyl-CoA : N-acetylneuraminate 4-O-acetyltransferase and the corresponding 7- or 9-O-acetyltransferase(s) have been discovered in equine and bovine submandibular glands. Acetylation of hydroxyl groups of N-acetylneuraminic acid has also been demonstrated in cell free systems.

Recently it has been shown that modification of the N-acetyl into the N-glycolyl compound, as well as O-acetylation, can occur by the action of a hydroxylase (which requires ascorbic acid and oxygen for its action) or of an O-acetyl transferase, respectively, after the NANA had been incorporated into the glycoprotein. This is another example of a modification reaction at the polymer level (Fig.36).

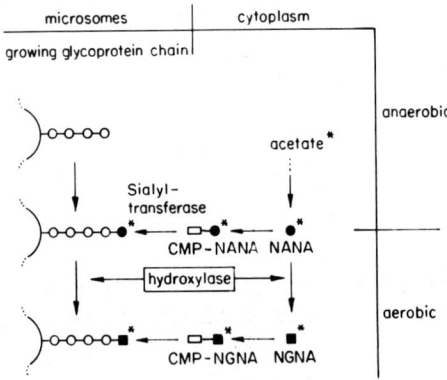

Figure 36 Hydroxylation of N-acetylneuraminic acid, either free or after incorporation into glycoprotein. NGNA, N-glycolylneuraminic acid. Modified from Schauer (2).

Activation of sialic acid occurs by its conversion into a nucleotide sugar, CMP-NANA, which is of unusual structure (Fig.37): it contains only one phosphate group, in contrast to all nucleotide-linked sugars we have encountered till now, which are diphosphate (or pyrophosphate) derivatives. Moreover, this sugar nucleotide is formed from NANA and cytidine triphosphate in a reaction

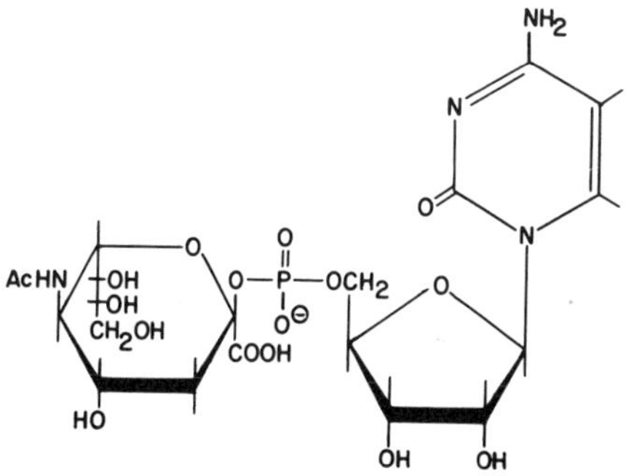

Figure 37 Cytidine monophosphate N-acetylneuraminic acid (CMP-NANA).

which is irreversible:

$$\text{NANA} + \text{CTP} \longrightarrow \text{CMP-NANA} + \text{PP}_i$$

All other reactions which lead to the formation of sugar nucleotides from the respective nucleoside triphosphates and sugar phosphates are reversible. It should also be noted that NANA is not phosphorylated prior to the reaction with CTP. To explain these unusual facts, it has been assumed that in the synthesis of CMP-NANA, the first step is the nucleophilic attack of the carboxyl of NANA on the α-phosphate of CTP, leading to the formation of a mixed anhydride (Fig.38). The mixed anhydride being rather unstable (or "energy rich") rearranges in an irreversible reaction to the final product, CMP-NANA in which the linkage between

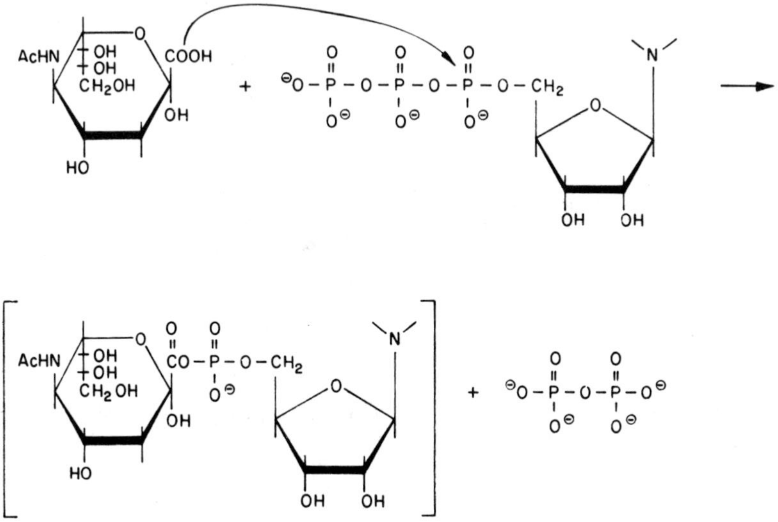

Figure 38   Hypothetical first step in the reaction leading to the formation of CMP-NANA from N-acetylneuraminic acid and cytidine triphosphate.

NANA and the phosphate is a β-glycosidic one. There is, however, no evidence for this mechanism.

The enzyme which synthesizes CMP-NANA is also known as CTP:acyl-neuraminate-cytidylyl-transferase. When isolated from different submandibular glands, it was found to be unspecific with regard to the various N- and O-acyl substituents of sialic acids.

As we shall soon see, the CMP-sialic acids are transferred to growing glycoprotein (and glycolipid) molecules by different sialyltransferases which exhibit specificity with regard to the acceptors of the sialic acids. However, the nature and position of the N- and O-acyl substituents of biologically occurring acylneuraminic acids appear to have no significant influence on the activity of the transferases.

The enzymes synthesizing N-acetylneuraminic acid from hexoses and converting it to CMP-NANA are known to occur in the cytoplasm. The enzymes modifying N-acetylneuraminic acid, and transferring the different sialic acids to glycoproteins appear to be firmly bound to subcellular membranes, probably to the Golgi apparatus.

REFERENCES

Reviews

1. The Chemistry and Biology of Sialic Acids and Related Substances,
   A. Gottschalk, Cambridge University Press, 1960, 115pp.
   Interesting and very readable.

2. Chemistry and biology of the acylneuraminic acids,
   R. Schauer, Angew. Chem. Intern. Ed. <u>12</u>, 127-138 (1973).
   Excellent and up-to-date review.

Specific articles

A1. New sialic acids. Identification of N-glycolyl-O-acetylneuraminic acids and N-acetyl-O-glycoloylneuraminic acids by improved methods for detection of N-acyl and O-acyl groups and by gas-liquid chromatography,
H. P. Buscher, J. Casals-Stenzal and R. Schauer, Europ. J. Biochem. <u>50</u>, 71-82 (1974).

A2. Configuration of the ketosidic bond of sialic acid,
R. K. Yu and R. Ledeen, J. Biol. Chem. <u>244</u>, 1306-1313 (1969).

A3. Studies on the mechanism of action of uridine diphosphate N-acetylglucosamine 2-epimerase,
W. L. Salo and H. G. Fletcher, Jr., Biochemistry 9, 882-885 (1970).

A4. Uridine diphosphate N-acetyl-D-glucosamine 2-epimerase from rat liver,
K. M. Sommar and D. B. Ellis, Biochim. Biophys. Acta 268, 590-595 (1972).

GLYCOPROTEIN BIOSYNTHESIS - II

BIOSYNTHESIS OF SACCHARIDE SIDE CHAINS IN CELL FREE SYSTEMS

The formation of the carbohydrate moieties of glycoproteins occurs by enzymic transfer of single sugar residues from nucleotide-linked sugars to nonreducing terminal positions of the growing saccharide side chains, as well as to amino acid side chains on proteins. Much information on these transfer reactions has accumulated as a result of extensive studies in cell free systems with isolated, though not always purified, enzymes (1).

Let us start by considering the relatively simple case of the biosynthesis of the carbohydrate side chains of collagen. In this protein either a single galactose unit or the disaccharide glucosylgalactose is linked to hydroxylysine. This amino acid is formed by the postribosomal hydroxylation of lysine side chains of the collagen precursor. Its hydroxyl group then serves as an acceptor of a galactose residue from the activated donor, UDP-Gal. This transfer reaction is catalysed by a galactosyltransferase which is specific to the donor and the protein acceptor. The second step in the biosynthesis is the transfer of a glucose unit from UDP-Glc to the galactose residue linked to the hydroxylysine by another glycosyltransferase of different specificity and cellular location. This glucosyltransferase will not attach glucose residues to hydroxylysine, nor to any acceptor other than to the protein-linked galactosyl-hydroxylysine. The specificity of enzymes is, however, not as rigid as that of a template, and their reactivity is influenced by environmental factors such as the presence of donors and acceptors. Therefore, the number of unsubstituted hydroxylysine residues in collagen and of those which carry galactose, or Glc-Gal, may vary. This variation in collagen is another example of the microheterogeneity, so characteristic of the carbohydrate side chains of glycoproteins.

The same sequence of reactions is involved in the synthesis of the glomerular basement membrane, also a collagen-type glycoprotein (Fig.39). Another group of

GLYCOPROTEIN BIOSYNTHESIS II

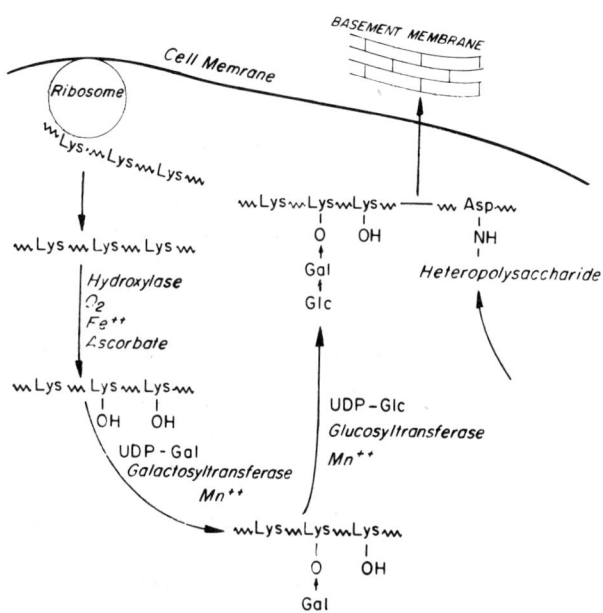

Figure 39 Schematic representation of some of the steps involved in the synthesis of the glomerular basement membrane (modified from Spiro (2)).

glycosyltransferases is responsible for the assembly of the asparagine-linked heteropolysaccharide units located on the more polar regions of the peptide chains of the basement membrane.

In the case of the plasma glycoproteins, α_1-acid glycoprotein and fetuin, the enzymic synthesis of the terminal trisaccharide unit (Fig.19) has been achieved in a cell free system with the aid of isolated glycosyltransferases.

Three steps are involved, according to the following sequence:

$$\text{UDP-GlcNAc} + \text{Man} \to \text{R} \longrightarrow \text{GlcNAc} \to \text{Man} \to \text{R} + \text{UDP} \quad (a)$$

$$\text{UDP-Gal} + \text{GlcNAc} \to \text{Man} \to \text{R} \longrightarrow \text{Gal} \to \text{GlcNAc} \to \text{Man} \to \text{R} + \text{UDP} \quad (b)$$

$$\text{CMP-NANA} + \text{Gal} \to \text{GlcNAc} \to \text{Man} \to \text{R} \longrightarrow$$
$$\text{NANA} \to \text{Gal} \to \text{GlcNAc} \to \text{Man} \to \text{R} + \text{CMP} \quad (c)$$

Reactions a, b and c are catalyzed by N-acetylglucosaminyl-, galactosyl and sialyltransferases, respectively. The physiological acceptor is usually a suitable glycoprotein or glycolipid (Man→R in the above sequence). In fact, some transferases have an absolute requirement for an acceptor of high molecular weight. However, certain glycosyltransferases can also transfer sugar to monosaccharides or small oligosaccharides, but it is often found that the small molecule is less efficient as acceptor. A transferase is usually specific for a particular sugar nucleotide and these enzymes are therefore conveniently classified according to the sugar transferred, i.e. sialyltransferases, galactosyltransferases, etc. No exception has yet been found to the rule that a single transferase catalyzes the synthesis of a single type of linkage. The acceptor specificity is usually determined by the sugar residue at the non-reducing end; in some instances, as will be discussed in more detail below, the penultimate sugar and its linkage to the terminal sugar are also important in determining acceptor activity.

An important feature of the reactions discussed is that the product of each glycosyltransferase reaction becomes the substrate for the next enzyme in the sequence, a property that S. Roseman from Johns Hopkins University has termed "cooperative sequential specificity". He has also introduced another concept, that of a multiglycosyltransferase system, or MGT (3), a complex of glycosyltransferases that catalyzes the synthesis of oligosaccharide side chains in glycoproteins (or gangliosides).

SPECIFICITY AND DISTRIBUTION

Pig liver has been shown to contain the four glycosyltransferases involved in the assembly of the two terminal trisaccharides of Asn-GlcNAc-type prosthetic groups, i.e. L-Fuc-Gal-β-(1→4)-GlcNAc and NANA-Gal-β-(1→4)-GlcNAc. Studies of substrate specificities of sialyltransferases from pig liver, rat mammary gland and goat colostrum have shown that the rat enzyme can be readily differentiated from the other two enzymes by its inability to transfer sialic acid to a high molecular weight acceptor. Also of great interest is the finding that goat colostrum and pig liver sialyltransferases can transfer sialic acid to N-acetyllactosamine (Gal-β-(1→4)-GlcNAc) but not to lactose or to the β-(1→3) and β-(1→6) isomers of N-acetyllactosamine; pig liver fucosyltransferase has a similar substrate specificity. The data indicate that a terminal galactosyl residue, a penultimate N-acetylglucosaminyl residue and a β-(1→4) linkage between the two are all essential components of the acceptor for both the sialyl- and fucosyltransferases.

The subcellular distribution of the glycosyltransferases involved in the assembly of the NANA-Gal-GlcNAc terminal trisaccharide of the GlcNAc-Asn-type carbohydrate units has been studied in rat liver. The Golgi-rich fraction is relatively enriched in the glycosyltransferase activities studied. Nuclei, mitochondria, rough-surfaced microsomes and post-microsomal supernatant have low levels of glycosyltransferase activity. These findings are in accord with the role of the Golgi apparatus in the biosynthesis of glycoproteins as described in lecture 7.

In this connection it is of particular interest that evidence has been obtained suggesting that glycosyltransferases may be present on the cell surface. These surface transferases may be involved in cell-cell interaction phenomena although no conclusive evidence for such a role has yet been obtained (3). It is possible that the surface enzymes were originally present in the Golgi apparatus and may have become externalized by the process of reverse pinocytosis, which is responsible for secretion of macromolecules from cells.

The intracellular glycosyltransferases are all membrane-bound and are activated by detergents such as Triton X-100. The degree of binding to membrane varies from one enzyme to the other, but all of the multiglycosyltransferase systems appear to be firmly anchored to the membrane; whereas solubilization is often readily achieved by the use of detergents, removal of membrane from the enzyme and subsequent purification are usually difficult procedures.

CONTROL OF GLYCOPROTEIN BIOSYNTHESIS

Only a limited amount of information is available on the many factors which control the synthesis of glycoproteins. Whereas genes control the assembly of activated amino acids into polypeptides by an accurate template mechanism, the synthesis of polysaccharide prosthetic groups is controlled by a non-template mechanism in which genes code for a large variety of glycosyltransferases.

Attachment of carbohydrate side chains is initiated

through the action of special glycosyltransferases capable of incorporating monosaccharides into side-chains of peptide-bound amino acids. Detailed studies have been carried out on the transferases responsible for synthesis of the GalNAc-Ser(Thr), Gal-Hyl and Xyl-Ser linkages; in all three cases, the transferases require a well-defined high molecular weight polypeptide as acceptor and thus initiation of prosthetic group synthesis appears to be controlled by the appropriate amino acid sequence and the acceptor specificity of the respective glycosyltransferase.

After initiation has occurred, elongation of the oligosaccharides is controlled primarily by the specificity of the MGT system for the acceptors. As mentioned, every transferase provides the substrate for the next transferase. Since the substrate specificites of glycosyltransferases are relative rather than absolute, the assembly of oligosaccharide prosthetic groups is subject to error. This phenomenon is probably one of several factors responsible for the occurrence of microheterogeneity in glycoproteins.

In addition to specificity, the action of the transferase depends on a variety of factors, such as the position of amino acid residues near the acceptor site, the ionic environment of the cell, and the presence of inhibitors or activators of the enzyme. It is therefore not surprising at all that the structures of the heterosaccharide side chains of glycoproteins do not show the same uniformity which we expect to find in the amino acid sequence of the protein, for the latter is the direct result of the DNA sequence.

The factors which terminate oligosaccharide chain elongation are unknown. Since sialic acid or fucose usually occupy terminal non-reducing positions in glycoproteins, it is possible that their attachment serves as a signal for chain termination.

The large number of different glycoproteins secreted by the liver suggests the presence within the liver cell of a correspondingly large number of multiglycosyltransferase systems. However, competition studies carried out with pig liver sialyltransferase indicate that a single transferase incorporates sialic acid into N-acetyllactosamine and into desialated α_1-acid glycoprotein and fetuin. Although many more such studies are needed, the data suggest that whenever the Gal-β-(1→4)-GlcNAc terminus is presented to the Golgi apparatus of the liver, a single sialyl-transferase attaches sialic acid to the galactose residue of the above mentioned disaccharide sequence. It is probable that the terminal trisaccharides of all the Asn-GlcNAc-type side chains are synthesized by a single MGT system in liver. Little is known, however, about the synthesis of the oligosaccharide cores, composed of N-acetylglucosamine and mannose residues. Elucidation of this problem must await the isolation and characterization of the glycosyltransferases responsible for synthesis of the core, which is in progress in a number of laboratories (see pp. 122-124).

The presence of at least one other MGT system in pig liver has recently been reported. Thus, pig liver contains a sialyltransferase with a specificity different from that of the enzyme acting on GlcNAc-Asn-type prosthetic groups. This sialyltransferase acts on the Ser(Thr)-GalNAc-type prosthetic groups of certain glycoproteins.

It is nevertheless very likely that many different glycoproteins share a common MGT system within the liver cell. Thus, for example, fucosyltransferase and sialyltransferase probably compete for Gal-β-(1$\rightarrow$4)-GlcNAc termini on a variety of different glycoproteins and it is not known how the sialic acid to fucose ratio is varied from one glycoprotein to another. Nor is it clear what mechanism controls the variations in linkage of sialic acid to galactose (2→2, 2→3, 2→4, and 2→6). A separate membrane assembly line for each glycoprotein would be one possible control mechanism; the MGT system of each assembly line would then be specially tailored for the synthesis of a particular glycoprotein prosthetic group. In such a model, the same glycosyltransferases may be present on many different assembly lines but the assembly lines would differ in the relative concentrations of various transferases and in their physical arrangement along the membrane.

The synthesis of submaxillary mucin is an interesting example of how two types of glycoproteins are produced on the same assembly line. Ovine submaxillary mucin (OSM) carries predominantly a disaccharide side chain prosthetic group, NANA-GalNAc (pp. 111-112). Porcine submaxillary mucin (PSM) has a more complex structure comprising the NANA-GalNAc disaccharide, to the GalNAc moiety of which are attached residues of galactose, $\underline{L}$-fucose and, in certain animals, N-acetylgalactosamine. Ovine submaxillary glands synthesize predominantly OSM-type mucin but also make a small amount of PSM-type mucin; porcine glands make a spectrum of compounds between the OSM-type mucin and the complete PSM-type mucin. A study of the glycosyltransferases involved in this process has suggested the probable control mechanism. The key enzyme is the galactosyltransferase which incorporates galactose into terminal peptide-linked N-acetylgalactosamine, provided the latter is not substituted by sialic acid; in other words, the incorporation of sialic acid prevents further incorporation of galactose. Ovine submaxillary glands have relatively high levels of the sialyltransferase and low levels of the galactosyltransferase, whereas the reverse is found in porcine submaxillary glands. Thus ovine glands make predominantly OSM-type mucin because sialic acid is incorporated into the growing mucin more readily than galactose. The reverse is true in porcine glands allowing synthesis of the more complex PSM-type mucin.

SPECIFICATION OF ANOMERY

In sugar nucleotides, the glycosidic linkage with D-sugars is always of the α-anomeric configuration, while with L-sugars (e.g. L-fucose) it is β. However, in glycoproteins, as in other sugar containing compounds, β-D and α-L linkages are also found. A case in point is mannose, derived from the α-linked GDP-Man, which as we have seen occurs in glycoproteins both as α and as β linked.

How are the two different types of linkages formed from the same sugar nucleotide? Following a proposal made by D. E. Koshland over 20 years ago, we assume that inversion of configuration, leading to the formation of a β-D-linkage, proceeds by a "single displacement mechanism". In such a reaction, the acceptor combines directly with the donor.

Retention of configuration is the result of a "double displacement mechanism", in which a covalent enzyme-bound intermediate is formed.

Covalent enzyme-substrate intermediate

In the covalent enzyme-substrate intermediate formed, the configuration of the anomeric linkage has been inverted. Subsequently, the acceptor reacts with the sugar linked to the enzyme, with a second inversion of configuration. Although no direct evidence for such a mechanism is available for reactions in which sugar nucleotides participate, covalently linked intermediates have been isolated in other reactions, where configuration of an anomeric bond is retained (e.g. with sucrose phosphorylase). Studies on the mechanism of action of lysozyme have, however, raised the possibility that retention of configuration may result even without formation of a covalent enzyme intermediate.

The above mechanisms are probably operating in the synthesis of other sugar polymers, e.g. glycogen (α-linked) and cellulose (β-linked), both of which are formed from α-linked sugar nucleotides: the former from UDP-Glc, the latter from GDP-Glc (or UDP-Glc as well) (4).

FEEDBACK CONTROL MECHANISMS IN GLYCOPROTEIN BIOSYNTHESIS

Two feedback controls have been demonstrated in the metabolic sequence leading to the formation of UDP-GlcNAc and of CMP-NANA, that serve as precursors for the biosynthesis of glycoproteins (A1) (Fig.40). Formation of UDP-GlcNAc is controlled at the level of the reaction leading to the synthesis of glucosamine-6-phosphate from fructose-6-P, since formation of the latter intermediate is common to many metabolic pathways.

Regulation of CMP-NANA levels and thus of glycoprotein

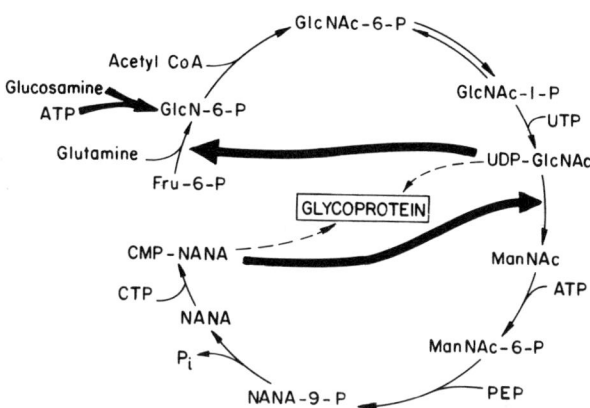

Figure 40 Pathways of biosynthesis of UDP-N-acetylglucosamine and CMP-N-acetylneuraminic acid in liver. The short dark arrow denotes the site of entry of injected glucosamine. The long dark arrows indicate sites of feedback inhibition (from Al).

biosynthesis, is possible by a feedback mechanism which consists of inhibition of the enzyme UDP-GlcNAc 2-epimerase by the above nucleotide sugar. This enzyme which catalyzes the formation of N-acetylmannosamine is the first that is specific to the sequence of reactions leading to the formation of CMP-NANA.

ROLE OF LIPID LINKED INTERMEDIATES

The mechanism of glycoprotein synthesis just discussed envisages the successive direct transfer of monosaccharides from corresponding sugar nucleotides to the growing oligosaccharide chain and, until very recently, was believed to be the only one operating in plants and animals. However, recent work from the laboratory of L.F.Leloir, as well as from other laboratories, including those of E. C. Heath, J.L.Strominger and R. W. Jeanloz, provides evidence that a mechanism analogous to that involved in the synthesis of polysaccharides in bacteria, i.e. via glycosylated lipid intermediates, may be functioning in plants and animals as well (5).

The transfer of sugars from sugar nucleotides to lipid acceptors in mammalian systems is now a well established fact. Beyond this fact, however, many questions still remain open. What is the chemical structure of the lipid acceptor? What is the mechanism of the transfer reaction and what are its products? What is the nature of the ultimate receptor for the lipid-bound sugar?

The nature of the lipid has not yet been conclusively established, but it is generally believed to be dolichol phosphate. Dolichol is the name of a family of polyprenols containing from 17 to 22 isoprene units, common to many mammalian tissues.

$$^{\ominus}O-\overset{O}{\underset{\underset{O}{\ominus}}{\overset{\|}{P}}}-O-CH_2-CH_2-\overset{CH_3}{\underset{|}{CH}}-CH_2-(CH_2-CH=\overset{CH_3}{\underset{|}{C}}-CH_2)_{15-20}-CH_2-CH=\overset{CH_3}{\underset{|}{C}}-CH_3$$

This conclusion was originally based on the fact that the acceptor lipid for glucose isolated from pig liver could be replaced by authentic dolichol phosphate, with which it was chromatographically identical. Subsequently, the utilization of exogenous dolichol phosphate as acceptor lipid has been confirmed in a number of other mammalian systems. Recently, the nature of the lipid has been determined by direct analysis of the endogenous acceptor. It was found to be a dihydropolyisoprenol, containing at least 18 isoprene units, one of which is saturated, and is thus a form of dolichol. It is not yet clear whether the same lipid or distinct members of a family of isoprenoid lipids function as carriers for different sugars.

The sugars that have been shown to participate in the transfer reaction to lipid are mannose, N-acetylglucosamine and glucose. Two types of linkages were found between the sugar and the lipid moiety. Whereas N-acetylglucosamine is transferred from UDP-GlcNAc as GlcNAc-1-P to give dolichol-PP-GlcNAc, both glucose and mannose are transferred from their respective nucleotides without phosphate to give dolichol-P-glucose and dolichol-P-mannose, respectively. In addition to the products described above, incubation of rat liver microsomes with UDP-GlcNAc as donor gave dolichol-PP-N,N'-diacetylchitobiose, while UDP-Glc and GDP-Man gave products identified as dolichol-PP-oligosacharides.

Similar findings have been reported with other systems that synthesize glycoproteins, such as mouse myeloma tumor (A2), human lymphocytes (A3) and calf pancreas (A4). On the basis of available results it seems probably that the following reaction sequence may take place (A5):

UDP-GlcNAc + dolichol-P → dolichol-PP-GlcNAc + UDP (a)

UDP-GlcNAc + dolichol-PP-GlcNAc →
  dolichol-PP-(GlcNAc)$_2$ + UDP (b)

GDP-Man and/or dolichol-P-Man + dolichol-PP-(GlcNAc)$_2$ →
  dolichol-PP-(GlcNAc)$_2$(Man)$_x$ (mannosylated
  endogenous acceptor) (c)

dolichol-P-Glc + dolichol-PP-(GlcNAc)$_2$(Man)$_x$ →
  dolichol-PP-(GlcNAc)$_2$(Man)$_x$(Glc)$_2$ (glucosylated
  endogenous acceptor) (d)

where (Man)$_x$ stands for mannose oligosaccharides ranging from 3 to about 16 units. However, it cannot be excluded that other sugars besides mannose are present.

Transfer of mannose to dolichol-PP-(GlcNAc)$_2$ (step c) to form mannosylated endogenous acceptor takes place either with GDP-Man or dolichol-P-Man as donors. One possibility is that dolichol-P-Man is a necessary intermediate in the transfer from GDP-Man; the other is that both compounds can act independently as donors.

The oligosaccharide moiety of mannosylated endogenous acceptor was transferred to a high molecular weight acceptor which showed many characteristics of a protein. The detailed nature of the acceptor has not yet been established.

Some consideration of the anomeric configuration of the compounds involved is of interest. The sugar nucleotides UDP-Glc, UDP-GlcNAc and GDP-Man are all α-linked, whereas dolichol-P-Man is β-linked (A6). It may be assumed that the transfer of the sugar moiety from the dolichol phosphate sugar occurs either with retention or with inversion of configuration, as is the case for sugar nucleotides in general. Transfer of N-acetylglucosamine (from dolichol-PP-GlcNAc, in which the sugar is presumably α-linked) to form, for example, the GlcNAc-β-(1→4)-GlcNAc sequence, and attachment of the bulk of the mannose residues that are α-linked will proceed with inversion of configuration. The single β-linked mannose residue, found in the core region of the carbohydrate units of many glycoproteins, seems to originate from GDP-Man (A7) and is also formed with inversion of configuration. Examples of retention of configuration in the formation of lipid-linked oligosaccharides have not been recorded.

The mannose oligosaccharide described above [i.e. $(GlcNAc)_2(Man)_x$] has a structure that is found in the "core" of many glycoproteins. It is therefore of particular interest that this moiety was transferred to a fraction that is most probably a protein. These results seem to indicate that the oligosaccharide moiety of a protein could be built up in an intermediate compound prior to transfer to protein.

It is still too early to try to integrate these findings into the accepted and well documented concept of glycoprotein synthesis by direct stepwise transfer of sugars from nucleotides. Location of the lipid in the membrane permits it to mediate the transfer of low molecular weight hydrophylic compounds that serve as building blocks of macromolecules localized within or beyond the hydrophobic cell membrane. It has therefore been suggested that the lipid-mediated mechanism will be found to operate in the synthesis of certain glycoproteins only (or of certain parts of the oligosaccharide side chain) depending on the nature and future cellular destination of the glycoprotein.

In this connection I cannot resist the temptation to tell you that in a review on polysaccharides that I wrote in 1965, just after the discovery of the lipid-linked oligosaccharides that serve as intermediates in the biosynthesis of cell wall glycopeptides and of bacterial cell wall peptidoglycan and lipopolysaccharides, I said (6):

" Similar lipid-linked compounds may function as intermediates in the biosynthesis of mucopolysaccharides and the carbohydrate moieties of glycoproteins."

REFERENCES

Reviews

1. The biosynthesis of animal glycoproteins,
    H. Schachter and L. Rodén *in* Metabolic Conjugation and Metabolic Hydrolysis (Ed. W. H. Fishman), Vol.3, Academic Press, 1973, pp.1-149.

2. Biochemistry of the renal glomerular basement membrane and its alterations in diabetes mellitus,
    R. G. Spiro, New England J. Med. 288, 1337-1342 (1973).

3. The synthesis of complex carbohydrates by multiglycosyltransferase systems and their potential function in intercellular adhesion,
    S. Roseman, Chem. Phys. Lipids 5, 270-297 (1970).
    A classic, with many interesting ideas; highly recommended.

4. Biosynthesis of oligosaccharides and polysaccharides in plants,
    W. Z. Hassid, Science 165, 137-144 (1969).
    Very readable and interesting article, written by one of the pioneers in this field.

5. The role of lipid-linked activated sugars in glycosylation reactions,
    W. J. Lennarz and M. G. Scher, Bioenergetics 4, 239-251 (1973).

6. Polysaccharides,
    N. Sharon, Ann. Rev. Biochem. 35, 485-520 (1966).

Specific articles

A1. The feedback control of sugar nucelotide biosynthesis in liver,
    S. Kornfeld, R. Kornfeld, E. F. Neufeld and P. J. O'Brien, Proc. Nat. Acad. Sci. USA 52, 371-379 (1964).

A2. The role of a dolichol-oligosaccharide as an intermediate in glycoprotein biosynthesis,
A. F. Hsu, J. W. Baynes and E. C. Heath, Proc. Nat. Acad. Sci. USA 71, 2391-2395 (1974).

A3. Transfer of sugars from nucleoside diphosphosugar compounds to endogenous and synthetic dolichyl phosphate in human lymphocytes,
J. F. Wedgwood, J. L. Strominger and C. D. Warren, J. Biol. Chem. 249, 6316-6324 (1974).

A4. Mannosyltransferase activity in calf pancreas microsomes. Formation from guanosine diphosphate-D-[^{14}C]mannose of a ^{14}C-labeled mannolipid with properties of dolichyl mannopyranosyl phosphate,
J. S. Tkacz, A. Herscovics, C. D. Warren and R. W. Jeanloz, J. Biol. Chem. 249, 6372-6381 (1974).

A5. Formation of lipid-bound oligosaccharides containing mannose. Their role in glycoprotein synthesis,
N. H. Behrens, H. Carminatti, R. J. Staneloni, L. F. Leloir and A. I. Cantarella, Proc. Nat. Acad. Sci. USA 70, 3390-3394 (1973).

A6. Occurrence of a β-D-mannopyranosyl phosphate residue in the polyprenyl mannosyl phosphate formed in calf pancreas microsomes and in human lymphocytes,
A. Herscovics, C. D. Warren, R. W. Jeanloz, J. F. Wedgwood, I. Y. Liu and J. L. Strominger, FEBS Letters 45, 312-317 (1974).

A7. Mannose transfer to lipid linked di-N-acetylchitobiose,
J. A. Levy, H. Carminatti, A. I. Cantarella, N. H. Behrens, L. F. Leloir and E. Tábora, Biochem. Biophys. Res. Commun. 60, 118-125 (1974).

10

FUNCTIONS OF THE CARBOHYDRATE - I

Until very recently the role of carbohydrate moieties in glycoproteins was almost completely unknown. With the rapid advances in our knowledge of the distribution, structure and metabolism of the saccharide units in glycoproteins, this role is slowly being elucidated (1). There is now an increasing number of glycoproteins for which we have good evidence or reasonable and sound proposals for the function of the sugar in their molecule. In many more cases it is becoming clear to us what properties the sugar residues, in particular the peripheral ones, impart to the intact glycoprotein molecule, and what changes the glycoprotein undergoes upon removal of these moieties. For most glycoproteins, however, the role of the carbohydrate units is still a subject for speculation.

At a meeting on glycoproteins in Lille in June 1973, H. Schachter proposed a classification of the functions of protein-bound carbohydrate. It is convenient to use this classification as a framework for our discussion.

POSTULATED FUNCTIONS OF PROTEIN-BOUND CARBOHYDRATES

(A) Physico-chemical properties of the glycoprotein, e.g. the viscosity of mucin

(B) Molecule-membrane interactions:
  (a) Secretion
  (b) Clearance of glycoproteins from plasma
  (c) Reaction of the cell surface with viruses, blood group antisera and lectins

(C) Membrane-membrane interactions:
  (a) Secretion
  (b) Differentiation and growth
  (c) Contact inhibition
  (d) Adhesion and aggregation of cells
  (e) Segregation of cells within the organism, e.g. the homing of lymphocytes
  (f) Gamete recognition

(A) PHYSICO-CHEMICAL PROPERTIES OF THE GLYCOPROTEIN. As mentioned in an earlier lecture (p. 48), attachment of sugars to proteins is known to increase the solubility of the latter. There are, however, other more remarkable effects of sugars on the physico-chemical properties of glycoproteins.

Acylneuraminic acids in glycoproteins are partly responsible for the high viscosity of mucilaginous secretions

of the respiratory tract, the urogenital tract and the eye
socket. Because of their low pK values, their carboxyl
groups are fully dissociated at physiological pH values.
The high density of the negatively charged carboxyl group
thus present on each mucin molecule, imparts to these molecules an extended rod-like structure. This rod-like charged
polyelectrolyte structure is responsible for the very high
viscosity of the aqueous solutions of sialic acid-rich
glycoproteins. Indeed, if we remove sialic acid from mucins
- either enzymically or by mild acid hydrolysis, their
viscosity will drop very markedly.

Mucilaginous glycoprotein solutions are vital for
animals. They act as lubricants for the rotation of
the eyeball, prevent the cornea from drying out and
protect it against damage by grains of dust. In the oral
cavity and in the gastrointestinal tract they envelop foods,
so making them slippery and protecting the tender mucous
membranes from mechanical damage. Furthermore, mucins protect the gastrointestinal tract from chemical agents, such
as stomach acid and digestive enzymes. In this way neuraminic acid is believed to oppose the formation of ulcers. The
layer of mucin that is continuously transported outward on
the epithelium of the respiratory passages provides moisture

for the incoming air, traps bacteria and other airborne impurities, some of which may be corrosive, and thus keeps the lung sterile. Similarly, a very viscous plug of mucin in the cervical canal of the uterus keeps bacteria out of the uterine cavity and hence out of the abdominal cavity. Most interestingly, this viscous barrier is lowered, by a mechanism not well understood, only at the time of ovulation to admit spermatozoa. Glycoproteins rich in neuraminic acid that are secreted by mucous glands of the vagina are important to reproduction in that they facilitate both coitus and childbirth.

Negatively charged acylneuraminic acid residues appear to impart a certain physical strength to cell membranes because of their mutual repulsion, and influence the mutual adhesion of cells in organ structure. Neuraminic acid present on the surface of blood platelets prevents their spontaneous clumping, and so opposes any undesirable formation of blood clots.

Sialic acid is not the only sugar which markedly affects the conformation of glycoproteins. According to R. Marshall, the major role of the carbohydrate moieties of some glycoproteins might be to assist in the maintenance of a specific tertiary structure. In this connection I would like to mention the studies of J. H. Pazur and his coworkers (A1) on the role of the carbohydrate moieties in glucamylase I from *Aspergillus niger*. This enzyme has a molecular weight of approximately 110,000 and contains 15% sugar by weight. Its saccharide side chains are linked by O-glycosidic bonds to approximately 45 serine and threonine residues of the polypeptide backbone. It was found that extensive oxidation by periodate of the carbohydrate residues in the enzyme markedly reduced its stability upon storage in the cold. It was suggested that by stabilizing the tri-dimensional structure of the glycoprotein, the carbohydrate moieties also affect the catalytic properties of the molecule.

Let me also remind you that in the anti-freeze glycoproteins (p. 39) the carbohydrate moiety is essential for activity. Thus, activity is abolished upon periodate oxidation or removal of the saccharide side chains (e.g. by alkali) or upon acetylation of one-fourth of the hydroxyl groups of the carbohydrate (A2).

We should, however, remember that with other glycoproteins there is evidence that the carbohydrate is not involved in the biological activity. I have already mentioned several times that ribonuclease B has the same enzymic activity as its carbohydrate-free counterpart, ribonuclease A. Work in our laboratory has very recently demonstrated (A3) that integrity of the carbohydrate chain is not required for hemagglutinating and mitogenic activities of soybean agglutinin: five out of the nine mannose residues attached to each subunit of this glycoprotein could be oxidized with periodate without affecting the above mentioned biological activities of soybean agglutinin. Also, the biological activity of interferon appears to be unaffected by the removal of sialic acid (A4).

Stability to proteolytic enzymes is not strictly a physical property; still it is important to keep in mind the many examples in which sugars are known to protect glycoproteins against proteolytic attack. Thus, α_1-acid glycoprotein is not attacked by trypsin, but becomes susceptible to this enzyme after removal of sialic acid. N-Acetylneuraminic acid protects the "intrinsic factor", a glycoprotein which binds B_{12} in the stomach, against the action of proteolytic enzymes. Other examples of this type are also known.

(B) SUGARS AND THE INTERACTION BETWEEN MACROMOLECULES AND MEMBRANES. A number of very important biological phenomena are the result of interactions of molecules with cells, in which carbohydrate moieties of glycoproteins are known to play a crucial role. The first of these to be recognized is the influenza virus hemagglutination phenomenon (p. 28). Much of our knowledge of this phenomenon is based on studies of the ability of various glycoproteins to inhibit hemagglutination by influenza virus. This activity is completely abolished upon treatment of the inhibitory glycoprotein with neuraminidase. However, inhibitory activity is affected by factors other than the presence of sialic acid. Thus, α_1-acid glycoprotein, containing 12% neuraminic acid, is only a weak inhibitor in comparison with the urinary Tamm and Horsfall glycoprotein which contains 7% neuraminic acid. The inhibitory activity is also affected by molecular size, since for example with α_1-acid glycoprotein, it has been found to increase after polymerization.

In order to be secreted, a protein synthesized in the cell must pass through the cell membrane. In 1966, E. Eylar of the University of Southern California proposed a general role for sugars in glycoproteins, that of "passport for transport" (2). This proposal was based on a survey of published data on the composition of over 100 proteins and glycoproteins, in which Eylar noted that most extracellular proteins were glycosylated, whereas intracellular ones were not, leading to the conclusion that the attachment of sugars represented a recognition signal to enable the cell to segragate intracellular and extracellular proteins.

This conclusion is no longer acceptable, in particular since it is clear that many secretions contain high proportions of non-glycosylated proteins. An extreme example is bovine pancreatic juice, in which only 4.7% of protein is in the form of glycoproteins (Fig. 3, p.35). We should, however, keep in mind the possibility that secreted non-glycosylated proteins were synthesized as glycoproteins and have lost their carbohydrate side chain when transported out of the cell.

The recent discovery of a disease in which collagen - normally a glycoprotein - is produced and secreted without any sugar attached to it, casts further doubt on Eylar's proposal, and makes it necessary for us to revise our basic concepts regarding how cells synthesize and secrete the proteins found in the extracellular spaces of the body (A5). The collagen in the skin of two sisters afflicted with the disease was found to be deficient in hydroxylysine, and as a result no sugar could be attached to the collagen. The defect is almost certainly a deficiency of the enzyme that hydroxylates the lysine residues in protocollagen. If such a collagen could not be secreted, the patient could not have survived fetal life, since the structural strength of skin and all other non-mineralized connective tissues depends almost entirely on extracellular collagen fibers.

This finding has caused much concern and efforts have been made to determine the exact amount of hydroxylysine in the skin collagen in the patients. It was found that their skin was not entirely free of hydroxylysine but contained about 0.2 residues per polypeptide chain. The collagen molecule contains three polypeptide chains, but the value of 0.2 per chain indicates that on the average each molecule would contain less than one residue of hydroxylysine. There are technical problems in assaying hydroxylysine at this low level, but if the values are accurate, we are left with only two alternatives. One is to say that the cells synthesizing collagen in most tissues in the two patients are altered somehow so that they can secrete a kind of collagen that normal cells cannot secrete. The other is to conclude that we have gone astray in our basic concepts of how cells synthesize and secrete collagen. If the concept of a sugar-tag is wrong for collagen, it may also be wrong for other extracellular proteins, and the process by which cells synthesize proteins for export is even more mysterious than it now seems.

In discussing the effect of saccharide moieties on molecule-membrane interaction, it is worthwhile to consider the immunological properties of glycoproteins. It has been noted that when animals are immunized with a glycoprotein such as fetuin, the elicited immune response is predominantly directed at the protein portion of the molecule and not at the carbohydrate moiety (A6). Investigation of the immunogenicity of glycopeptides derived from human immunoglobulins have shown that they are very poor antigens (A7) and it was concluded that they contribute little to the antigenic determinants of the parent molecule. A study of chicken ovalbumin failed to reveal any evidence for the formation of carbohydrate-specific antibodies to this glycoprotein (A8).

The difficulty in raising antibodies to sugar portions of glycoproteins may be due to the obvious similarities in many of the carbohydrate structures appearing in otherwise unrelated serum glycoproteins, such as fetuin, thyroglobulin, Ig molecules of different classes, or cell surface glycopeptides including the histocompatibility antigens. Their similarities may have rendered animals tolerant to these structures, which would be recognizable as "self" antigens.

In contrast to most glycoproteins, the A,B,H(O) blood group substances are good antigens, the antibodies formed being specific for their carbohydrate moieties. As is well known, most of us carry the natural antibody against the

A, B or AB blood type determinants which are sugar specific. Another example of a glycoprotein where the sugar appears to be involved in immunological specificity is the glomerular basement membrane. Anti-glomerular basement membrane auto-antibodies play an important role in the pathogenesis of some cases of human glomerule nephritis. However, until recently, the chemical structure of basement membrane antigens responsible for this auto-immunization was a matter of controversy. It has now been shown (A9) that the major antigenic determinant of the glomerular basement membrane is a glycopeptide containing the disaccharide Glc-Gal. The glycopeptide was isolated from a proteolytic digest of the glomerular basement membrane and its amino acid sequence established as Hyl-Gly-Glu-Asp-Gly. The disaccharide prosthetic group is linked to the hydroxylysine residues of the peptide chain. The immunodominant antigenic site is formed by the Gal-Hyl unit.

In passing, let us not forget that many "pure" polysaccharides (e.g. the dextrans and the pneumococcal polysaccharides) are good antigens.

Recent work has shown that receptors interacting with mitogenic reagents such as lectins and sodium periodate to induce lymphocyte transformation, and in some cases immunoglobulin secretion, have structures resembling the carbohydrate side chains of certain glycoproteins. Another important phenomenon which falls under the heading molecule-membrane interaction is the clearance of glycoproteins from plasma. Both these phenomena will be the subject of the following lecture.

(C) CARBOHYDRATES IN MEMBRANE-MEMBRANE INTERACTIONS. A number of observations have implicated surface membranes, and more specifically the glycoproteins or glycolipids of the cell surface, in the regulation of biological phenomena. In the first lecture I already told you (pp.28-29) that the fate of lymphocytes in circulation is greatly affected by removal of L-fucose from their surface. It has also been shown that certain monosaccharides, e.g. L-fucose and mannose, when added to cultures of fibroblasts, are able to bring about phenomena resembling contact inhibition of mitosis. There is suggestive evidence that carbohydrates may be involved in morphogenesis (A10): teratoma cells when grown singly will not aggregate unless glutamine (a precursor of glucosamine), glucosamine or mannosamine is added to the medium.

Changes in sugars on cell surfaces occur when normal cells are transformed into malignant ones (3,4). Such transformation also results in loss of contact inhibition, and this has been taken as evidence for the involvement of sugars in cellular recognition and intercellular communication. Tumor specific differences in cell surfaces can now be demonstrated through the use of lectins, since many lectins will agglutinate transformed cells at concentrations much lower than those required for the agglutination of the normal parental cells.

The results of M. C. Glick and L. Warren at the University of Pennsylvania, have revealed differences in the distribution of fucose-containing glycopeptides after transformation by RNA or DNA viruses (5). Thus far, all cells examined after viral transformation show a similar distribution of glycopeptides, which differs from the distribution found in membranes obtained from the normal counterparts of these cells.

Sugars on cell surfaces vary also throughout the cell cycle: when a normal cell is arrested in metaphase, the distribution of surface glycopeptides is more like that from the surface of cells transformed by oncogenic viruses than that from normal cells. The exact relationship of these fluctuations to cell division and viral transformation is not understood and is being investigated.

S. Roseman has proposed that intercellular adhesions are a result of mutual interactions between oligosaccharide units in the membrane of one cell with glycosyltransferases present in the membrane of the adjacent cells (6). In this mechanism, a glycosyltransferase on the surface of one cell may not only be involved in the synthesis of surface oligosaccharides, but may also be able to bind to the appropriate saccharide acceptor on an adjacent cell. This process would result in the adherence of the two cells; the specificity of the mutual adhesive recognition would depend on the high degree of specificity of the enzyme for the particular acceptor, as well as for the donor molecule. According to the suggested mechanism, on transfer of the appropriate saccharide residue from the sugar nucleotide donor to the acceptor on the adjacent cell surface, the enzyme-substrate complex dissociates and the cells separate. The transferase acceptor model is particularly attractive because by completing a glycosyltransferase reaction, and thus causing a small change in surface chemistry, the cells become detached.

A mechanism of this type could explain a variety of other surface phenomena, such as contact inhibition. Work aimed at providing evidence for or against this mechanism is currently in progress in several laboratories. (See also pp. 160-161).

REFERENCES

Reviews

1. Functional importance of surface heterosaccharides in cellular behaviour,
   G. M. W. Cook and R. W. Stoddart in Surface Carbohydrates of the Eukaryotic Cell, Academic Press, 1973, pp.257-293.

2. On the biological role of glycoproteins,
   E. H. Eylar, J. Theoret. Biol. 10, 89-113 (1965).

3. Surface changes in transformed cells detected by lectins,
   M. M. Burger, Fed. Proc. 32, 91-101 (1973).

4. Lectins: cell-agglutinating and sugar-specific proteins,
   N. Sharon and H. Lis, Science 177, 949-959 (1972).

5. Isolation and characterization of surface membrane glycoproteins from mammalian cells,
   M. C. Glick, Meth. Membrane Biol. 2, 157-204 (1974).

6. The synthesis of complex carbohydrates by multiglycosyltransferase systems and their potential function in intercellular adhesion,
   S. Roseman, Chem. Phys. Lipids 5, 270-297 (1970).

Specific articles

A1. Glycoenzymes: enzymes of glycoprotein structure,
    J. H. Pazur and N. N. Aronson, Jr., Adv. Carb. Chem. Biochem. 27, 301-341 (1972).

A2. Structure and role of carbohydrate in freezing point-depressing glycoproteins from an antarctic fish,
    J. R. Vandenheede, A. I. Ahmed and R. E. Feeney, J. Biol. Chem. 247, 7885-7889 (1972).

A3. Labeling of soybean agglutinin by oxidation with sodium periodate followed by reduction with sodium [^{3}H]borohydride,
    R. Lotan, H. Debray, M. Cacan, R. Cacan and N. Sharon, J. Biol. Chem. 250, 1955-1957 (1975).

A4. Interferon: evidence for its glycoprotein nature,
F. Dorner, M. Scriba and R. Weil, Proc. Natl. Acad.
Sci. USA 70, 1981-1985 (1973).

A5. A subtle disease and a dilemma: can cells secrete
collagen that does not contain a sugar-tag?
D. J. Prockop, New England J. Med. 286, 1055-1056 (1972).

A6. Fetuin: immunochemistry and quantitative estimation
in serum,
F. H. Bergmann, L. Levine and R. G. Spiro, Biochim.
Biophys. Acta 58, 41-51 (1962).

A7. Glycopeptides of human immunoglobulins - II. Contribution to the antigenicity of the heavy chain,
F. Miller, Immunochem. 8, 99-111 (1971).

A8. Investigation of the antigenicity of the carbohydrate
moiety of chicken ovalbumin,
M. Kaplan and M. Schlamowitz, Immunochemistry 9,
737-747 (1972).

A9. Primary structure of a small glycopeptide isolated
from human glomerular basement membrane and carrying
a major antigenic site,
P. Mahieu, P. H. Lambert and G. R. Maghuin-Rogister,
Eur. J. Biochem. 40, 599-606 (1973).

A10. An L-glutamine requirement for intercellular adhesion,
S. B. Oppenheimer, M. Edidin, C. W. Orr and S. Roseman,
Proc. Natl. Acad. Sci. USA 63, 1395-1402 (1969).

FUNCTIONS OF THE CARBOHYDRATE - II

SUGARS AND THE SURVIVAL OF GLYCOPROTEINS IN CIRCULATION

One of the most exciting developments in the study of glycoproteins has been the discovery of the role of carbohydrates on the clearance and survival of glycoproteins in circulating blood, and their uptake by the liver (1). This work, done by G. Ashwell from the National Institutes of Health together with A. Morell from Albert Einstein College of Medicine in New York, has led to the hypothesis that sialic acid is essential for the prolonged viability of many, if not most plasma glycoproteins in the circulation. Even partial desialylation results in prompt clearance of the modified protein. Since the sequence NANA-Gal-GlcNAc is of very common occurrence in plasma glycoproteins, removal of the terminal sialic acid exposes the subterminal galactose. The exposed galactose appears to be a necessary (but not sufficient) precondition for prompt catabolism of glycoproteins in plasma. It has further been observed that for uptake of the catabolised proteins by the liver, the presence of intact sialic acid residues on the receptor sites of the hepatic plasma membrane is an absolute requirement for the initial binding reaction which in turn is a prelude to transport into the cell and to lysosomal catabolism.

As often happens, these important discoveries originated from an unexpected observation made in 1966, during studies on the biological role of ceruloplasmin, a copper transport glycoprotein found in serum. For these studies, ceruloplasmin was radioactively labelled at its galactose moieties in the following manner (Fig.41): the terminal sialic acid residues were removed from ceruloplasmin by neuraminidase, unmasking the subterminal galactose residues; the latter were oxidized by galactose oxidase, and the resultant aldehyde derivative was reduced with tritiated borohydride.

Very surprisingly, when this radioactive asialoceruloplasmin was injected into rabbits, the label disappeared rapidly from circulation; 15 minutes after injection less

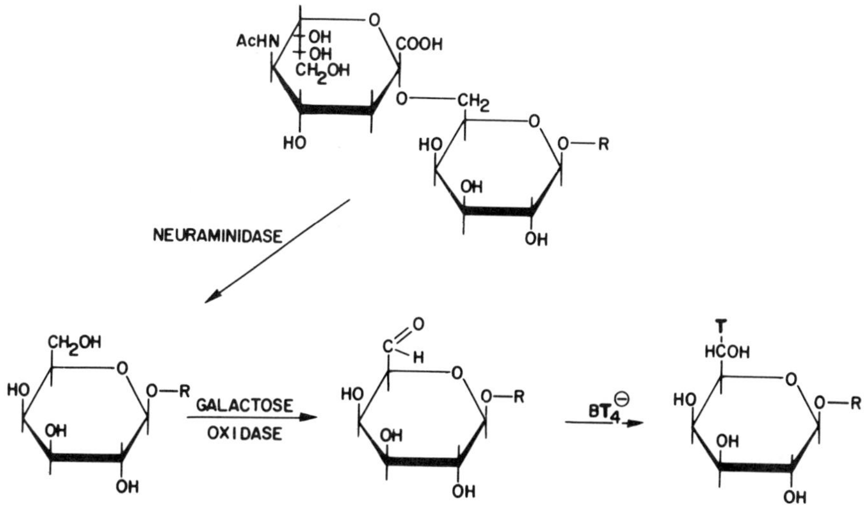

Figure 41 Radioactive labeling of penultimate galactosyl residues of glycoproteins

than 10% of asialoceruloplasmin was present in the serum, in striking contrast to native ceruloplasmin, over 90% of which remained in circulation after the same period of time (Fig.42). In other words, the half-life of ceruloplasmin in rabbits dropped dramatically from 56 hours to 3-5 minutes when the sialic acid had been removed from it. The same result was obtained irrespective of whether human or rabbit ceruloplasmin was used. Experiments with other serum glycoproteins such as fetuin, α_1-acid glycoprotein (Fig.42) and haptoglobin, have also shown that selective removal of sialic acid from their carbohydrate side chains leads to a very remarkable increase in their rate of disappearance from circulation.

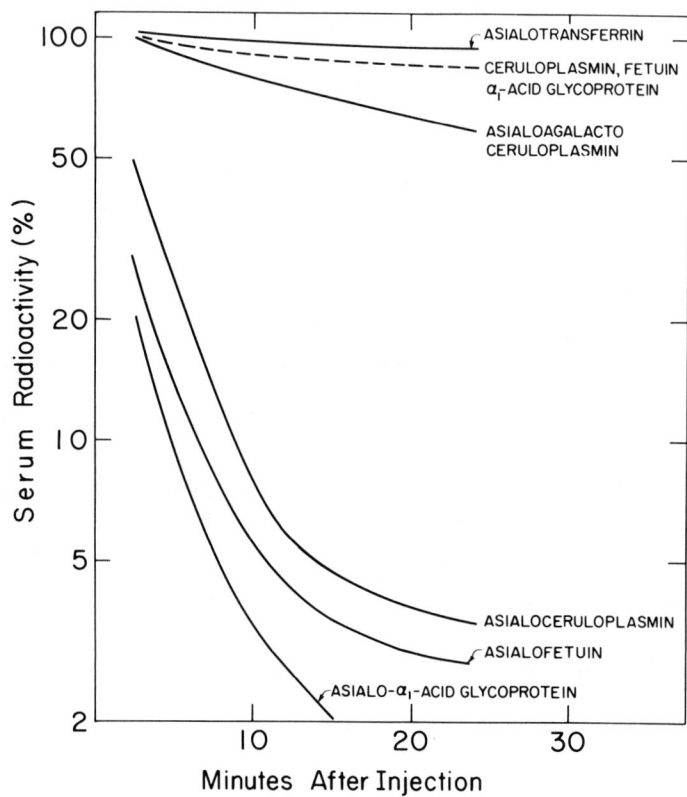

Figure 42 Disappearance from rabbit serum of radioactively labelled native and modified glycoproteins. Redrawn from the data of Ashwell and Morell (1).

This was, however, not true for all serum glycoproteins, since removal of sialic acid from transferrin, which also possesses the sequence NANA-Gal-GlcNAc, did not lead to a marked decrease in its survival in circulation (Fig.42).

Further studies have shown that soon after their disappearance from the serum, essentially all the injected sialic acid-free glycoproteins were recovered intact from the liver. It was also found that subsequent to their attachment to the liver plasma membrane, the asialoglycoproteins were transferred to the lysosomes, where the full machinery for the degradation of the peptide and carbohydrate moieties of glycoproteins resides. In rabbits from which the liver was taken out, asialoglycoproteins remained for a prolonged time in circulation.

The rapid removal from circulation of the asialoglycoproteins was shown to depend on the presence of non-reducing unmodified terminal galactose residues on their molecules. If this sugar, which had become exposed by the initial neuraminidase treatment, was oxidized by galactose oxidase or was taken off by β-galactosidase digestion, an almost complete restoration of the prolonged survival time of the native glycoproteins was obtained (Fig.42). These and other observations clearly show that intact terminal galactose residues are required for recognition by the plasma membrane of the liver cells.

How many sialic acid residues should be removed before the liver will recognize a deficient molecule? It was found that removal of only two out of the ten sialic acid residues present in a molecule of ceruloplasmin (molecular weight about 130,000) will mark it for destruction. Moreover, using neuraminidases that differ in their specificity for the position of the N-acetylneuraminyl linkage, it was shown that it did not matter which two residues of sialic acid were removed. Convincing proof that no other changes in the glycoprotein were involved was provided by demonstrating that the replacement of the missing sialic acid residues afforded a glycoprotein which was again normally viable in circulation. This was done with a sialyltransferase, which catalyses the attachment of sialic acid residues onto terminal galactose residues of glycoproteins, by transfer from CMP-NANA.

Ashwell and Morell concluded at this point that "exposure of any two galactosyl residues of ceruloplasmin is sufficient cause for the prompt removal of the 'defective' molecule from plasma".

I have mentioned earlier that all asialoglycoproteins that disappear rapidly from circulation are taken up by the liver. It was further shown that the liver has a limited capacity for the uptake of glycoproteins. Thus, when unlabelled asialo-α_1-acid glycoprotein was injected into a rat together with labelled asialoceruloplasmin, the survival of the latter in circulation was markedly increased. The increase in survival time was dependent on the ratio of the two asialoglycoproteins injected: the higher the proportion of the asialo-α_1-acid glycoprotein, the slower was the rate of clearance from circulation of the labelled asialoceruloplasmin. Fully sialylated glycoproteins, such as intact α_1-acid glycoprotein, were ineffective in lengthening the survival of tracer amounts of asialoceruloplasmin, nor were glycopeptides isolated from pronase digests of such glycoproteins. However, desialylated glycopeptides from ceruloplasmin or α_1-acid glycoprotein inhibited the uptake of asialoceruloplasmin by the liver; no inhibition was observed with simple saccharides such as galactose or lactose. These and other experiments have led to the conclusion that the information for uptake by the liver is contained mainly, if not entirely, in the oligosaccharide moiety of the glycoproteins. It is not known, however, which factors besides the terminal non-reducing galactose are involved in the uptake phenomenon.

Removal of sialic acid from glycoprotein hormones and from erythrocytes also affects dramatically their rate of clearance from circulation. Rabbit erythrocytes have a half-life in circulation of 9-11 days. After enzymic hydrolysis of sialic acid from their surface, their half-life drops to as low as 7-12 hours (A1,A2). Since old erythrocytes have a lower content of sialic acid than young ones, it is possible that removal of sialic acid by neuraminidase is involved in the physiological aging of erythrocytes. Such a mechanism may also be responsible for the elimination by the body of old platelets, leucocytes and other types of old cells.

Some fifteen years ago it was found that the glycoprotein hormones human chorionic gonadotropin (HCG) and follicle stimulating hormone (FSH) from which the sialic acid was removed, were biologically inactive. This led to the belief that the integrity of the carbohydrate side chain of glycoprotein hormones was essential for their biological activity. As standard hormone assays are done with intact animals, the possibility was raised that the loss of the potency of the desialylated HCG and FSH might be the result of their rapid removal from the circulation and destruction by the liver. Ashwell and Morell have, indeed, demonstrated that radioactively labelled desialylated HCG and FSH disappeared rapidly from circulation in rats and were recovered from the liver. However, when asialo-α_1-acid glycoprotein was injected into the animals simultaneously with the desialylated hormones, not only did their survival time in circulation increase, but they proved to be biologically active. This has also been found with other glycoprotein hormones, such as desialylated luteinizing hormone (LH) and erythropoietin. With the development of *in vitro* assays for hormones it was confirmed that the biological activity of the glycoprotein hormones was not affected by changing their sialic acid content. It was further found that for binding to crude hormone receptor fractions prepared from testis and ovary, sialic acid in HCG and LH is completely dispensable.

In the course of these studies, a new method for the labelling of sialoglycoproteins was developed. According to this method (Fig.43) the two distal exocyclic carbon atoms of sialic acid (C-8 and C-9) are selectively cleaved by periodic acid under very mild conditions (0°C, 10-15 min), which do not affect any other parts of the glycoprotein

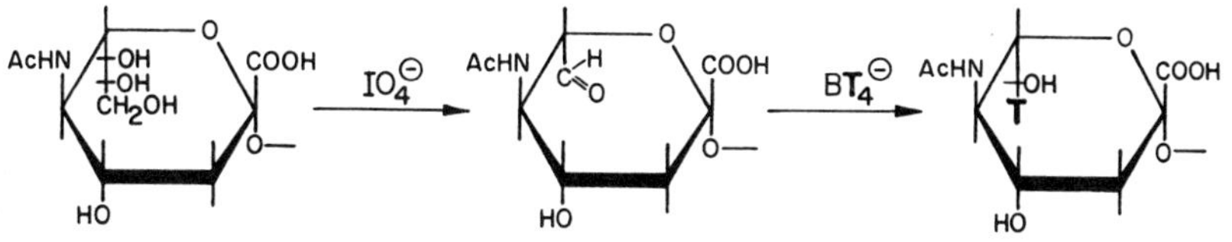

Figure 43 Radioactive labelling of terminal sialic acid residues of glycoproteins (A3).

molecule. The resulting C-7 aldehyde is reduced with tritiated borohydride to yield a stable radioactive derivative of the original glycoprotein (A3). All the sialic acid is thus converted into its 7-carbon analog, 5-acetamido-3,5-dideoxy-L-arabino-2-heptulosonic acid (NANA-7 or Neu-7). Upon injection into rats, preparations of ceruloplasmin and α_1-acid glycoprotein thus labelled exhibited a normal half-life, establishing that the integrity of sialic acid is not essential for the survival of the glycoproteins in circulation. The 7-analog of sialic acid could be removed from the labelled glycoproteins by neuraminidase, although at a rate slower than native sialic acid, affording desialylated glycoproteins which were rapidly cleared from circulation. Using the same technique, modification of the sialic acid residues of the glycoprotein hormones HCG and FSH into their Neu-7 analogs was achieved with little loss of hormonal activity (A4). The method is also applicable to the labelling of the surface of erythrocytes and lymphocytes, and is finding an extensive use in the study of sugars on cell membranes (A5). Incidentally, cell surface glycoproteins and glycolipids can also be labelled at their galactose residues following removal of sialic acid (A6), employing the same method as that used for the labelling of soluble glycoproteins (see Fig.41).

The importance of the carbohydrate in directing the uptake of plasma proteins by liver cells was emphasized by the work of J. C. Rogers and S. Kornfeld, who showed in 1971 (A7) that clearance of serum albumin (a protein devoid of carbohydrate) from the circulation in rats was stimulated at least 30-fold by coupling it chemically with a fetuin glycopeptide from which the sialic acid had been removed. As expected from the studies of G. Ashwell and A. Morell, no stimulation of hepatic uptake of serum albumin was brought about by coupling it with the intact fetuin glycopeptide which contained sialic acid residues. This work of Rogers and Kornfeld provides a method for controlling the lifetime of specific proteins injected into circulation, and for directing them to the liver and ultimately to the hepatic lysosomes. It has far reaching implications as a potential means for replacement therapy in cases of genetic deficiency of lysosomal and other enzymes.

It is also obvious from these and other studies that the catabolism of serum albumin, which does not contain carbohydrate, must be directed by a mechanism quite different from that just described. Alternate pathways of catabolism probably also exist for some glycoproteins, since, as mentioned, the clearance from serum of transferrin does not seem to be enhanced by removal of its sialic acid (Fig.42).

LIVER MEMBRANE RECEPTORS. I have mentioned earlier that the plasma membranes of the liver have been implicated as the locus of the binding sites for circulating sialic acid-free glycoproteins. Using isolated liver membranes, the binding of asialoglycoproteins could be quantitated and striking differences between different asialoglycoproteins were observed. For example, binding of asialo-α_1-acid glycoprotein was on a molar basis about 100 times stronger than that of asialoceruloplasmin. Here, as in the experiments with intact animals, terminal galactose residues were shown to be the major determinants of binding, but no insight was obtained for the other factors in the asialoglycoprotein which are responsible for the broad spectrum of binding observed.

The binding of asialoglycoproteins to liver membranes was specific, as no binding to crude membrane preparations from other organs was detected.

Effective binding of asialoglycoproteins by isolated liver plasma membranes was shown to depend on the presence of calcium ions and on the sialic acid residues of the membranes. Incubation of the membranes with extremely small amounts of neuraminidase promptly and completely abolished their ability to bind asialoglycoproteins. Reconstruction experiments in which membranes inactivated by partial desialylation were incubated with CMP-NANA, led to restoration of a significant portion of the lost binding activity. Here no addition of sialyltransferase was necessary, since this enzyme was present in the membrane fraction. With the demonstration that the neuraminidase-mediated loss of binding potency of the plasma membranes was a repairable lesion, it became clear that sialic acid plays a complex and mutually exclusive role in determining the metabolic fate of specific circulating glycoproteins. Effective hepatic uptake requires not only the <u>absence</u> of sialic acid on the glycoprotein, but in addition the <u>presence</u> of this substituent on the plasma membranes.

All the binding activity of the liver membrane for asialoglycoproteins and asialoglycopeptides was recently shown to reside in one of its glycoprotein constituents. This glycoprotein was isolated, first from liver membrane preparations, and subsequently from a more convenient source - an acetone powder of rabbit liver (A8, A9). Purification of the hepatic binding glycoprotein was achieved by affinity chromatography on a column made of asialo-α_1-acid glycoprotein covalently attached to Sepharose. Binding to the column was in the presence of Ca^{++}, and the purified glycoprotein was eluted simply by passing through it a suitable buffer solution free of Ca^{++}. The purified hepatic binding glycoprotein was polydisperse; it contained 10% carbohydrate identified as N-acetylneuraminic acid, galactose, mannose and glucosamine in the molar ratios 1:1:2:2. Its binding properties were exactly as could be predicted from the experiments with the liver membranes. For example, no inhibition of the binding of asialo-α_1-acid glycoprotein was affected even by very high concentrations of intact α_1-acid glycoprotein or ceruloplasmin. Binding required Ca^{++} and the presence of sialic acid on the hepatic glycoprotein. Very interestingly, the unmodified hepatic binding glycoprotein, but not its desialylated derivative, agglutinated human and rabbit erythrocytes. Hemagglutination was specifically inhibited by monosaccharides, such as N-acetylgalactosamine and galactose, and by asialo-α_1-acid glycoprotein. Carbohydrate-binding proteins exhibiting such agglutinating activity, known as lectins, have been isolated from a variety of plants, invertebrates and lower vertebrates. The hepatic-binding glycoprotein is the first lectin of mammalian origin.

To summarize, it is conceivable that glycoprotein

catabolism may represent an example of a finely poised system regulated by the relative activities of the two complementary enzymes, neuraminidase and sialyltransferase, acting in concert upon both the surface carbohydrate of the liver cells and the circulating glycoproteins of the blood. It is also reasonable to assume that the role of surface carbohydrates in influencing glycoprotein catabolism represents a process of major physiological significance. Such an inference must, however, be made with caution since direct experimental confirmation of this biological function is not yet available.

CELL SURFACE SACCHARIDES AND LYMPHOCYTE ACTIVATION

In my lectures I have emphasised several times the important role that sugars on cell surfaces play in the life of the cell. In an interesting application of the techniques employed by Ashwell and Morell for labelling glycoproteins, A. Novogrodsky and E. Katchalski in our department have demonstrated that chemical modification of sugars on the surface of lymphocytes has dramatic effects on these cells: specific oxidation of cell surface sialic acid or galactose stimulated lymphocytes to undergo metabolic and morphological changes culminating in growth and cell division. The effects observed, as measured for example by the stimulation of DNA synthesis, were comparable with those induced by well known mitogens, such as lectins (phytohemagglutinin and Con A) that bind to cell surface saccharides (Fig.44), or by antibodies to cell surface components.

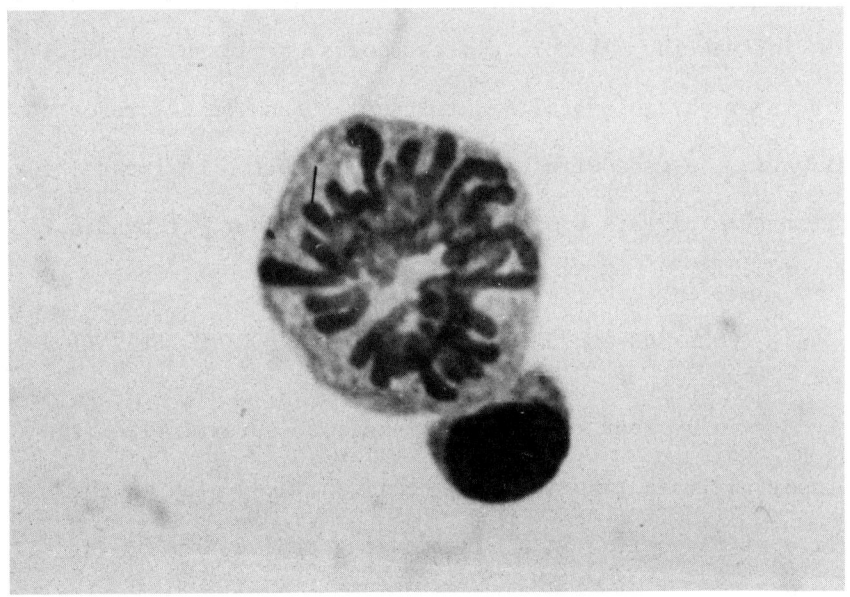

Figure 44 Micrograph (∼1000x) of a mouse spleen lymphocyte undergoing mitosis after 3 days incubation in the presence of concanavalin A (2 µg/ml). An unstimulated cell is seen next to a stimulated one.(Courtesy A. Novogrodsky).

Studies of mitogen induced lymphocyte transformation are of extreme importance, since they serve as an excellent model for investigating a large number of biological phenomena. These include the mechanism of transmission of messages from the cell surface to its interior, the cellular events which cause lymphocytes to become active in the immune system, for example in producing antibodies, and the relation between structure and function of membranes. The discovery that lymphocytes can be stimulated by oxidation offers new and exciting possibilities for the study of the above problems, since this modification is specific, well defined chemically and can be readily controlled.

Novogrodsky has shown that lymphocytes from a variety of sources (e.g. man, mouse or rat), will undergo mitogenic stimulation upon mild treatment with periodate, under conditions which affect only sialic acid residues on their surface (A10). Treatment of lymphocytes with neuraminidase markedly reduced their response to periodate. Moreover, the triggering signal generated by periodate oxidation could be abolished by treatment with reagents that modify aldehyde groups, either by reduction (e.g. KBH_4) or by condensation (e.g. hydroxylamine).

Transformation of lymphocytes was also induced by aldehyde groups formed on their surface by sequential treatment of the cells, first with neuraminidase and then with galactose oxidase (A11). No stimulation was observed when galactose oxidase was incubated with cells from which the sialic acid had not been removed, nor did neuraminidase itself act as a mitogen. Moreover, no mitogenic effect was observed when the neuraminidase treated lymphocytes were incubated with galactose oxidase in the presence of a specific inhibitor of the enzyme, or an excess of galactose (substrate inhibition). The above findings clearly show that in this case stimulation is the result of oxidation of the galactose residues on the lymphocyte surface. As in the case of the aldehyde group on oxidized sialic acid, modification of the aldehyde group formed by neuraminidase and galactose oxidase treatment markedly decreased the mitogenic effect of the sequential enzymic treatment.

The mode of action of the aldehyde groups formed on the lymphocyte surface in stimulating lymphocyte proliferation is not known. Aggregation of specific membrane sites is believed to be involved in the stimulation of lymphocytes by mitogens and other triggering agents. It is possible that the aldehydes on C-7 of the modified sialic acid or C-6 of galactose react with amines or other functional groups on the cell surface forming crosslinks; these crosslinks lead to the formation of cell surface aggregates which act in the triggering process.

Interestingly, the response of lymphocytes treated

with periodate, or with neuraminidase and galactose oxidase, to stimulation by Con A, is the same as that of untreated cells, suggesting that different saccharide sites are involved. Indeed, it is believed that Con A induces transformation of lymphocytes by binding to mannose residues on the cell surface (2). Further evidence on the role of cell surface sugars in mitogenic stimulation has been obtained from studies with soybean agglutinin, a lectin which has been investigated in our laboratory for many years. This lectin, specific for terminal non-reducing N-acetylgalactosamine or galactose residues, is not mitogenic for rat, mouse or human lymphocytes. However, soybean agglutinin stimulated these lymphocytes after sialic acid had been removed from their surface (Al2). It was postulated that the transformation induced by soybean agglutinin in desialylated lymphocytes is caused by the binding of the lectin to newly exposed galactosyl residues. Since sialic acid in glycoproteins is known to be glycosidically linked to galactose (or N-acetylgalactosamine), Novogrodsky has further suggested that periodate, galactose oxidase and soybean agglutinin trigger lymphocytes to undergo transformation by affecting the same glycoprotein that contains one or more oligosaccharides with the sequence NANA-Gal (or GalNAc). No experimental evidence for this assumption is available, but attempts to isolate the putative glycoprotein are in progress.

Another interesting related finding made by Novogrodsky is that specific oxidation of sialic acid or galactose (after neuraminidase treatment) on the surface lymphocytes will convert them into killer cells that will lyse and destroy certain "target cells", for example mastocytoma cells (Al3). Also, treatment of target cells with periodate or neuraminidase-galactose oxidase renders them susceptible to cytolysis by untreated lymphocytes. These effects are also abolished by aldehyde modifying reagents, indicating that the presence of aldehyde groups on either the lymphocytes or the target cells is sufficient to induce cytotoxicity. It is possible that here, too, the aldehyde groups on the treated cells interact with amines or other chemical groups on the surface of the untreated cells, to form intercellular bridges, with the resultant formation of a cytotoxic complex which leads to lysis of the target cells.

REFERENCES

Reviews

1. The role of surface carbohydrates in the hepatic recognition and transport of circulating glycoproteins,
    G. Ashwell and A. G. Morell, Adv. Enzymol. 41, 99-128 (1974).
    A comprehensive and very readable account of the subject, from which most of the material for this lecture was taken.

2. Lectins: cell-agglutinating and sugar-specific proteins,
    N. Sharon and H. Lis, Science 177, 949-959 (1972).

Specific articles

A1. The role of sialic acid in the determination of survival of rabbit erythrocytes in the circulation,
    L. Gattegno, D. Bladier and P. Cornillot, Carbohyd. Res. 34, 361-369 (1974).

A2. Sialic acid - a determinant of the life-time of rabbit erythrocytes,
    J. Jancik and R. Schauer, Hoppe-Seyler's Z. physiol. Chem. 355, 395-400 (1974).

A3. Studies on the chemical and enzymatic modification of glycoproteins,
    L. Van Lenten and G. Ashwell, J. Biol. Chem. 246, 1889-1894 (1971).

A4. Effect of modification of N-acetylneuraminic acid on the biological activity of human and ovine follicle-stimulating hormone,
    M. Suttajit, L. E. Reichert, Jr. and R. J. Winzler, J. Biol. Chem. 246, 3405-3408 (1971).

A5. Modification of sialyl residues of sialoglycoprotein(s) of the human erythrocyte surface,
    T. H. Liao, P. M. Gallop and O. O. Blumenfeld, J. Biol. Chem. 248, 8247-8253 (1973).

A6. External labeling of cell surface galactose and galactosamine in glycolipid and glycoprotein of human erythrocytes,
C. G. Gahmberg and S. I. Hakomori, J. Biol. Chem. 248, 4311-4317 (1973).

A7. Hepatic uptake of proteins coupled to fetuin glycopeptide.
J. C. Rogers and S. Kornfeld, Biochem. Biophys. Res. Commun. 45, 622-629 (1971).

A8. The isolation and properties of a rabbit liver binding protein specific for asialoglycoproteins,
R. L. Hudgin, W. E. Pricer, Jr. and G. Ashwell, J. Biol. Chem. 249, 5536-5543 (1974).

A9. Mammalian hepatic lectin,
R. J. Stockert, A. G. Morell, and I. H. Scheinberg, Science 186, 365-366 (1974).

A10. Membrane site modified on induction of the transformation of lymphocytes by periodate,
A. Novogrodsky and E. Katchalski, Proc. Nat. Acad. Sci. USA 69, 3207-3210 (1972).

A11. Induction of lymphocyte transformation by sequential treatment with neuraminidase and galactose oxidase,
A. Novogrodsky and E. Katchalski, Proc. Nat. Acad. Sci. USA 70, 1824-1827 (1973).

A12. Transformation of neuraminidase-treated lymphocytes by soybean agglutinin,
A. Novogrodsky and E. Katchalski, Proc. Nat. Acad. Sci. USA 70, 2515-2518 (1973).

A13. Induction of lymphocyte cytotoxicity by modification of the effector or target cells with periodate or with neuraminidase and galactose oxidase,
A. Novogrodsky, J. Immunology 114, 1089-1093 (1975).

BLOOD GROUP SUBSTANCES

- I: CHEMICAL STRUCTURE

The surface of the human erythrocyte, as that of other types of cell, is coated with a complex mosaic of a large number of specificity determinants, many of which are complex saccharides.

To date, over 250 red cell antigens have been described, and the number is rapidly increasing. Included in this number are some 100 blood group determinants, belonging to 15 independent human blood group systems. Of these, the most thoroughly studied are the antigenic determinants of the ABO blood group system, and the closely related Lewis system (1-4). Considerable information has also been obtained in recent years on the antigens of the MN system (4). Little is known, however, about the structure of other red cell antigens, such as those of the rhesus (Rh) blood group system.

The ABO blood group system, the first to be described, was discovered by K. Landsteiner in 1900 as a result of his attempts to determine whether specific serological differences existed between individuals of the same species. The importance of the knowledge of the ABO groups for the safe practice of blood transfusion is now widely recognized. All other human blood group systems known at present, except for the Rh system discovered some 40 years ago, are of little clinical significance.

The various blood groups are defined by their serological properties. That is, the type specific antigens are identified by means of suitable antibodies. Such antibodies may be derived from the same species in which the antigen is present, as is the case with the ABO system: the serum of humans of blood group A contains anti-B antibodies (anti-B isoagglutinins), and that of group B contains anti-A isoagglutinins (Table 12). More often (e.g. with the MN antigens), the antibodies are obtained by injection of erythrocytes of one species into another species.

The simplest way to demonstrate the presence of an antigen on erythrocytes is by testing whether they will be agglutinated by the specific antibodies. Lectins can also be used for this purpose, since some of them are blood type

Table 12

Relation between genotype, antigens on red cells, and antibodies in serum in ABO blood-group system

Group (phenotype)	Genotype	Antigen on red cell	Antibodies in serum
A	*AA* *AO*	A	anti-B
B	*BB* *BO*	B	anti-A
AB	*AB*	A and B	–
O	*OO*		anti-A and anti-B

specific (5). The lectins of *Phaseolus limensis* (lima bean), *Vicia cracca* and *Dolichos biflorus* are blood group A specific; *Lotus tetragonolobus*, *Ulex europeus* and the eel, are O (or H, see below) specific; *Iris amara* is M specific and *Vicia graminea* is N specific. Some of these are used in blood banks.

In this lecture I shall discuss mainly the antigenic determinants of the ABO and Lewis systems, which have been well characterized from a chemical standpoint and for which the biochemical basis of inheritance has recently been established.

The study of these determinants has provided us with

a beautiful example, perhaps the best of its kind, of the way in which oligosaccharides of cell surfaces are assembled, of how the pattern of that assembly is genetically determined, and of the great complexity and variability of structure that is produced.

In addition to their biological and medical significance, knowledge of blood group antigens is also useful for legal, and even historic purposes. The latter is possible because the information contained in complex carbohydrate polymers which are resistant to climatic and bacteriological destruction can be used in paleontology. In a dramatic example, the tissue dust obtained from the remains of Tutankhamen has been used in serological experiments to establish his blood group and his probable relationship to Smenkhkare, another eighteenth dynasty Pharaoh.

THE ABH AND LEWIS ANTIGENS

Blood group characters are inherited according to simple Mendelian laws, and the antigens are believed to be the products of allelic or closely linked genes. The A and B phenotypes are controlled by two allelic genes, A and B (Table 12). There is now ample evidence that, contrary to earlier beliefs, there is no O gene product and that the substance on group O cells, detected by the so called "anti-O" reagents (such as the O-specific lectins mentioned above) is a product of a gene system designated as Hh. This conclusion is based, among other things, on the finding that people who belong to blood group O, as well as those who belong to blood groups A, B or AB, often secrete a glycoprotein that inhibits agglutination of O erythrocytes by "anti-O" reagents. To avoid confusion, this glycoprotein is now called H substance, instead of O substance. The reagents that react with it, as well as with the corresponding H determinants, are anti-H reagents. For historical reasons, however, the designation "group O erythrocytes" has been retained.

Substances with ABH and Lewis activities commonly occur not only on erythrocytes but also on other cell membranes and in many secretions. Not all humans, however, have ABH antigens on their erythrocytes: a very small number of individuals lacks these antigens. The first cases belonging to this rare blood type were discovered in Bombay, India. Since such "Bombay" type individuals are apparently normal in all other respects, it would appear that the ABH antigens do not have a specific function essential for cell vitality.

The earliest indication that ABH specificity may be determined by sugars was obtained by W. M. Watkins and W. T. J. Morgan at the Lister Institute, London, in 1952. Of all sugar constituents of blood group substances, only N-acetylgalactosamine and its α-methyl glycoside inhibited agglutination of red cells by type A-specific lectins. Watkins and Morgan concluded, therefore, that this sugar, in α linkage, is a determinant of blood group A specificity. Similarly, agglutination of group O cells by the type H(O)-specific lectins was inhibited best by methyl α-L-fucopyranoside, indicating that the α-L-fucosyl residue is a determinant of H(O) specificity. In other words, α-linked N-acetylgalactosamine and L-fucose are the immunodeterminant sugars of blood type A and H(O) specificity, respectively. These conclusions have been fully substantiated in subsequent studies. Unfortunately, lectins were not useful for studying blood group B determinants, as blood group B-specific lectins were not known at the time.

Strangely enough, the detailed knowledge of the chemical structure of the ABH and Lewis (Le^a and Le^b) antigens is based mostly not on studies with red cells, but on soluble blood group active substances that are found in secretions such as saliva, ovarian cyst fluid or gastric juice. Isolation of these antigens from erythrocytes has not been possible until recently, because of the difficulty in obtaining reasonable amounts of pure materials. ABH specific glycolipids have now been isolated from human erythrocyte membranes but their investigation has been hampered by low yields and by problems of purification and polymorphism (4).

The magnitude of the problem will be appreciated from the fact that S. I. Hakomori obtained only 1 or 2 mg of a family of pure H-active glycosphingolipids from the stroma of 30 liters of blood. Whether erythrocytes contain ABH-active glycoproteins is still questionable. It should be noted, however, that the MN blood group antigens are part of the major human erythrocyte glycoprotein, also known as glycophorin.

The chemical structure of the ABH and Lewis specific blood group substances was elucidated primarily in the 1950's and 1960's by Morgan and Watkins in London and by E. A. Kabat and his coworkers at Columbia University College of Physicians and Surgeons in New York. The substances that have been studied most are those isolated from ovarian cyst fluids. These fluids accumulate within the cyst over long periods, and a single cyst sometimes yields several grams of active material. Purified soluble blood group substances are polydisperse materials, with molecular weights ranging from 3×10^5 to 10^6. Preparations from even a single individual and a single type of secretion, contain a family of molecules very closely related in general structure and composition. Such substances, irrespective of their blood group specificity, usually contain about 85% of carbohydrate and 15% protein, and are therefore glycoproteins. In the protein, hydroxyamino acids comprise a significant proportion (40-50% serine and threonine), while the proportion of aromatic and sulfur containing amino acids is low. The same sugars are present in preparations of different specificity: galactose, N-acetylglucosamine, N-acetylgalactosamine, L-fucose and a small (and variable) amount of sialic acid.

Because of the extreme complexity of the soluble blood group substances, and the fact that only a small portion of these macromolecules is responsible for their blood group activity, much effort had to be invested in elucidating their structure, a task far from complete. Many approaches, both direct and indirect, have been used for this purpose.

Enzymic degradation of blood group glycoproteins provided evidence for the role of L-fucose in H and Le[a] specificity, and of N-acetylgalactosamine and galactose in A and B specificity, respectively (Fig.45).

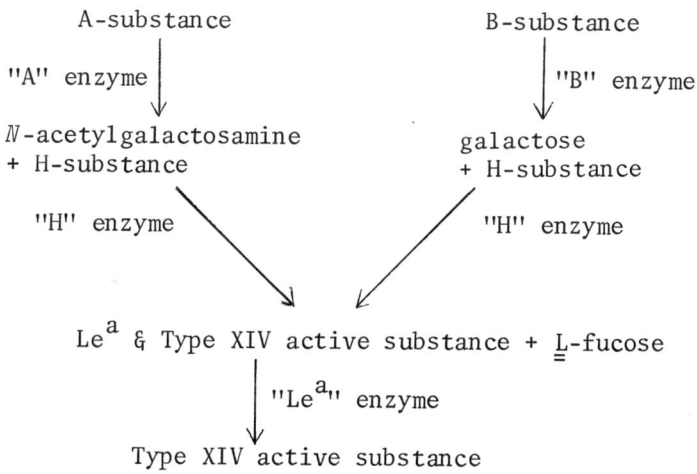

Figure 45 Specificities revealed by the sequential enzymic degradation of blood group substances A and B (from ref.1).

Originally, crude enzymes were used and specificity of action was based on inhibition of degradation by monosaccharides (substrate inhibition). Thus, Morgan and Watkins found that enzymes from the protozoon *Trichomonas foetus*, which destroyed A and B specificities with concomitant appearance of H activity, were inhibited by N-acetylgalactosamine and galactose, respectively. With the availability in recent years of purified glycosidases, it was confirmed that α-N-acetylgalactosaminidase will convert type A glycoprotein into H glycoprotein. Similarly, α-galactosidase converts B glycoprotein into H glycoprotein. Treatment of the residual H-active glycoprotein with a purified α-$\underline{L}$-fucosidase, results in the liberation of $\underline{L}$-fucose and the development or enhancement of cross reactivity with anti-Le^a antibody and also with horse antiserum to Type XIV pneumococcus, an antiserum specific for β-galactosyl residues. It appears, therefore, that structures related to H, Le^a, and Type XIV pneumococcus determinants are present in A and B glycoproteins, and that, since the release of a single sugar unit suffices to unmask a new specificity, the structures revealed are originally part of the same carbohydrate chains. Purified α-N-acetylgalactosaminidase and α-galactosidase can also be used to convert group A and B erythrocytes, respectively, to group O erythrocytes.

Much structural information was obtained by examining the ability of a variety of oligosaccharides to inhibit specifically hemagglutination or precipitation of blood group glycoproteins by their immune antisera ("hapten inhibition" experiments). The earlier hapten inhibition studies were done with oligosaccharides isolated from human milk and characterized structurally in the 1950's by R. Kuhn and his associates in Germany (for examples, see Table 13). These oligosaccharides are present in milk in small quantities, and can all be considered as derivatives of lactose, the

Table 13

Some oligosaccharides of human milk

Trivial name	Structure
Lactose	Gal-β-(1→4)-Glc

2'Fucosyllactose	Gal-β-(1→4)-Glc
	    2
	    ↑
	L-Fuc-α-1

3-Fucosyllactose	Gal-β-(1→4)-Glc
	           3
	           ↑
	       L-Fuc-α-1

Lacto-difucotetraose	Gal-β-(1→4)-Glc
	  2 3
	  ↑ ↑
	L-Fuc-α-1 L-Fuc-α-1

Lacto-N-tetraose	Gal-β-(1→3)-GlcNAc-β-(1→3)-Gal-β-(1→4)-Glc
Lacto-N-neotetraose	Gal-β-(1→4)-GlcNAc-β-(1→3)-Gal-β-(1→4)-Glc

Lacto-N-fucopentaose I	Gal-β-(1→3)-GlcNAc-β-(1→3)-Gal-β-(1→4)-Glc
	  2
	  ↑
	L-Fuc-α-1

Lacto-N-fucopentaose II	Gal-β-(1→3)-GlcNAc-β-(1→3)-Gal-β-(1→4)-Glc
	             4
	             ↑
	      L-Fuc-α-1

Lacto-N-fucopentaose III	Gal-β-(1→4)-GlcNAc-β-(1→3)-Gal-β-(1→4)-Glc
	             3
	             ↑
	      L-Fuc-α-1

Lacto-N-difucohexaose I	Gal-β-(1→3)-GlcNAc-β-(1→3)-Gal-β-(1→4)-Glc
	  2 4
	  ↑ ↑
	L-Fuc-α-1 L-Fuc-α-1

Lacto-N-difucohexaose II	Gal-β-(1→3)-GlcNAc-β-(1→3)-Gal-β-(1→4)-Glc
	       4 3
	       ↑ ↑
	  L-Fuc-α-1 L-Fuc-α-1

Lacto-N-hexaose	Gal-β-(1→4)-GlcNAc-β-1
	                 ↓
	                 6
	Gal-β-(1→3)-GlcNAc-β-(1→3)-Gal-β-(1→4)-Glc

principal milk sugar. They were of particular aid in providing information about the Lewis determinants. Inhibition of Le^a hemagglutination by lacto-N-fucopentaose II and of Le^b hemagglutination by lacto-N-difucohexaose I was observed, which suggested that the Le^a and Le^b glycoproteins differ in the number of L-fucose residues that serve as specificity determinants - one L-fucose for Le^a and two for Le^b.

The isolation and identification of more than 50 oligosaccharides (di- to decasaccharides) from the products of partial acid hydrolysis and alkaline degradation of blood group active glycoproteins, supplied the details necessary to interpret the complex inter-relationship of specificities revealed by the hapten inhibition and enzyme degradation experiments. All these compounds were, of course, tested for their serological activity. Some of the oligosaccharides proved to be excellent inhibitors in hemagglutination or precipitation experiments, indicating that they contain a large part of the antigenic determinant of each substance (A1,A2).

Among the A active fragments were the trisaccharides of GalNAc-α-(1→3)-Gal-β-(1→3)-GlcNAc and GalNAc-α-(1→3)-Gal-β-(1→4)-GlcNAc. Similar trisaccharides, with galactose instead of N-acetylgalactosamine at their non-reducing end, were isolated from B active glycoproteins, and inhibited the precipitation of these proteins with their antibodies. As can be seen in both the A and B active trisaccharides, the subterminal galactose is linked to the terminal reducing end by either 1→3 or 1→4 linkages. It was thus concluded that A, B and H specificity involves two types of basic structure, 1→3 called "Type 1 grouping" and 1→4 called "Type 2 grouping". The isolation by E. A. Kabat and his coworkers of branched oligosaccharides in which both types of grouping are present, clearly demonstrated that Type 1 and Type 2 groupings may occur as part of the same carbohydrate chain. The most complex oligosaccharide of this series was a decasaccharide, 4.4 mg of which was obtained from 4.56 g of a Le^a active glycoprotein which had been treated with NaOH-NaBH$_4$ (A1). The structure of this compound is given below, with the Type 1 and Type 2 groupings underlined.

$$
\begin{array}{c}
\underline{\text{L-Fuc-}\alpha\text{-1}} \\
\downarrow \\
4 \\
\underline{\text{Gal-}\beta\text{-(1→3)-GlcNAc-}\beta\text{-(1→3)}} \searrow \\
 \text{Gal-}\beta\text{-(1→3)-GlcNAc-}\beta\text{-(1→3)-} \\
 \nearrow \text{Gal-}\beta\text{-(1→3)-}N\text{-acetylgalac-} \\
 \text{tosaminitol} \\
\underline{\text{Gal-}\beta\text{-(1→4)-GlcNAc-}\beta\text{-(1→6)}} \nearrow \\
3 \\
\uparrow \\
\underline{\text{L-Fuc-}\alpha\text{-1}}
\end{array}
$$

As expected, it does not contain the A and B specific sugars (α-GalNAc and α-Gal, respectively) at its non-reducing end. The N-acetylgalactosaminitol present at the other end of this fragment has been produced by reduction with $NaBH_4$ of the N-acetylgalactosamine released by β-elimination from the carbohydrate-peptide linkage (see pp. 72-75).

A glycoprotein with oligosaccharide chains terminating with either Gal-β-(1→3)-GlcNAc or the Gal-β-(1→4)-GlcNAc sequences, has been termed a "precursor substance"; such a substance has been isolated from human ovarian cyst fluid.

Based on the different lines of evidence that I have summarized, it is clear that the ABH and Lewis glycoproteins possess a common basic structure, and that their blood group specificity is determined by the nature and linkage of the monosaccharides at the non-reducing ends of their carbohydrate chains. There are two kinds of non-reducing ends, Type 1 and Type 2, with galactose linked to N-acetylglucosamine either by a β-(1→3) or by a β-(1→4) linkage, respectively. To these structures are attached N-acetylgalactosamine, galactose, and L-fucose, to form determinants of A, B, H, Le^a and Le^b specificity, as shown in Table 14.

The A, B, and H determinants can arise from either Type 1 chains or Type 2 chains; Le^a and Le^b determinants can

Table 14

Structures responsible for A, B, H, Le^a and Le^b specificities

Specificity	Structure	
	Type 1	Type 2

H	Gal-β-(1→3)-GlcNAc..
	    2
	    ↑
	L̲-Fuc-α-1
		Gal-β-(1→4)-GlcNAc..
		    2
		    ↑
		L̲-Fuc-α-1

A	GalNAc-α-(1→3)-Gal-β-(1→3)-GlcNAc..
	                        2
	                        ↑
	                 L̲-Fuc-α-1
		GalNAc-α-(1→3)-Gal-β-(1→4)-GlcNAc..
		                        2
		                        ↑
		                 L̲-Fuc-α-1

B	Gal-α-(1→3)-Gal-β-(1→3)-GlcNAc..
	                 2
	                 ↑
	          L̲-Fuc-α-1
		Gal-α-(1→3)-Gal-β-(1→4)-GlcNAc..
		                 2
		                 ↑
		          L̲-Fuc-α-1

Le^a	Gal-β-(1→3)-GlcNAc..
	                4
	                ↑
	         L̲-Fuc-α-1	

Le^b	Gal-β-(1→3)-GlcNAc..
	    2 4
	    ↑ ↑
	L̲-Fuc-α-1 L̲-Fuc-α-1	

only be formed from Type 1 chains, since both structures contain L̲-Fuc-α-(1→4)-GlcNAc, a sequence that cannot be formed in Type 2 chains where the 4 position of N-acetylglucosamine is occupied by galactose.

As mentioned, the most important sugars for each specificity are known as immunodeterminants or immunodominant sugars. The immunodominant sugar for H specificity is L̲-fucose as L̲-Fuc-α-(1→2)-Gal; for A specificity, N-acetylgalactosamine as GalNAc-α-(1→3)-Gal; for B specificity, galactose as Gal-α-(1→3)-Gal; and for Le^a specificity, fucose as:

$$\begin{array}{c}\ldots\text{GlcNAc}\ldots\\ 4\\ \uparrow\\ \underline{\text{L}}\text{-Fuc-}\alpha\text{-1}\end{array}$$

Both fucose residues, as

$$\begin{array}{cc}\text{Gal-}\beta\text{-(1}\rightarrow\text{3)-GlcNAc}\ldots\\ 2\qquad\qquad 4\\ \uparrow\qquad\qquad\uparrow\\ \underline{\text{L}}\text{-Fuc-}\alpha\text{-1}\quad\underline{\text{L}}\text{-Fuc-}\alpha\text{-1}\end{array}$$

are required for Le^b specificity.

The difference in A and B specificity, of such great importance in blood grouping and transfusion, appears thus to reside ultimately in a relatively small structural variation, namely in the nature of the substituent at C-2 in a sugar with a galactose configuration.

Since both Type 1 and Type 2 groupings are common to all soluble blood group substances regardless of their A, B, H or Lewis specificity, these substances cross-react to varying degrees with horse anti-Type XIV pneumococcal serum, which is directed mainly against the Gal-β-(1→4)-GlcNAc (Type 2) grouping. Some 250-350 chains, many of them apparently branched, are believed to be attached to the same polypeptide by α-glycosidic bonds between the proximal N-acetylgalactosamine residue and the hydroxyl groups of serine or threonine. This linkage is alkali labile, which made possible the isolation, for example, of the decasaccharide described on page 227. There is considerable heterogeneity among the chains, however, and chains at various stages of completion and with variable fucose content have been isolated. This is another example of the microheterogeneity of glycoproteins, which we discussed in detail in an earlier lecture. It explains why single glycoprotein molecules possess different specificities: glycoproteins from A or B individuals, for example, may have variable amounts of H, Le^a or Le^b determinants in addition to A or B and Type XIV pneumococcal determinants. Precipitation experiments with specific antisera indicate that the various determinants are present on the same molecule. As to the structure of the polypeptide backbone of the blood group active glycoprotein, very little is known.

In passing, brief mention should be made of the glycolipid antigens with ABH and Lewis specificities (4). These have been investigated mainly by T. Yamakawa in Japan, S. I. Hakomori and his coworkers in Japan and the U.S., and J. Koscielak and his group in Poland. Unfortunately, no good method exists for the separation of glycolipids from mixtures with similar structures. The difficulty increases as the carbohydrate chain length and complexity of the molecule increase. Moreover, blood group active glycolipids are only very minor constituents of the erythrocyte membrane, and are not found in large quantities elsewhere in the body. In spite of these enormous difficulties, it has been recently established that they occur as families of compounds which differ in their glucosamine content. From blood type O erythrocytes, a glycolipid fraction was obtained showing inhibition of hemagglutination caused by the H-specific lectin of *Ulex europeus* and the precipitation of Le^b and H-specific glycoproteins by their immune antisera. Several H-active glycolipids have been isolated from this fraction. The structure of one of these is as follows:

$$\text{Gal-}\beta\text{-}(1\to4)\text{-GlcNAc-}\beta\text{-}(1\to3)\text{-Gal-}\beta\text{-}(1\to4)\text{-Glc-ceramide}$$
$$\underset{\underline{\text{L}}\text{-Fuc-}\alpha\text{-1}}{\overset{2}{\uparrow}}$$

In these compounds, the glucose is glycosidically linked to the terminal hydroxyl of ceramide (*N*-acylsphingosine) which has the following structure:

$$\text{HO-CH}_2\text{-}\underset{\underset{\underset{O}{\overset{\|}{\text{NH--C·R}}}}{|}}{\text{CH}}\text{-}\overset{\overset{\text{OH}}{|}}{\text{CH}}\text{-CH=CH-(CH}_2)_{12}\text{-CH}_3$$

The *N*-acyl group can be one of several different fatty acids.

Similar glycolipids, with A and B activities, were isolated from A and B erythrocytes, respectively. As expected, the A active glycolipids contained *N*-acetylgalactosamine linked α-(1→3) to the terminal non-reducing galactose of the H glycolipid, whereas the B active glycolipid contained a Gal-α-(1→3) grouping instead. It is interesting that the blood group active glycolipids contain glucose, and that they have the Type 2 rather than Type 1 grouping.

Characterization of Lewis glycolipids from human erythrocytes had not been done until recently. Le^a and Le^b active glycolipids are present in serum, and there is evidence that the Lewis antigens are acquired from the plasma by adsorption.

REFERENCES

Reviews

1. Blood group substances,
 W. M. Watkins, Science 152, 172-181 (1966).
 A basic review; though written some ten years ago, still very worthwhile reading.

2. Enzymatic basis for blood groups in man,
 V. Ginsburg, Advanc. Enzymol. 36, 131-149 (1972).
 Includes a brief and excellent discussion of the chemistry of the ABO blood group substances.

3. Blood group specific substances,
 W. M. Watkins *in* Glycoproteins (Ed. A. Gottschalk), 2nd edit., Part B, Elsevier, 1972, pp.830-891.
 For those interested in a more detailed survey on the subject.

4. Blood group antigens,
   S. I. Hakomori and A. Kobata *in* The Antigens (Ed. M. Sela), Vol.2, Academic Press, 1974, pp.79-140.

5. Lectins: cell-agglutinating and sugar-specific proteins,
   N. Sharon and H. Lis, Science $\underline{177}$, 949-959 (1972).

Specific articles

A1. Structures of oligosaccharides produced by base-borohydride degradation of human ovarian cyst blood group H, Le^b and Le^a active glycoproteins,
   L. Rovis, B. Anderson, E. A. Kabat, F. Gruezo and J. Liao, Biochemistry $\underline{12}$, 5340-5354 (1973).

A2. Activity of reduced oligosaccharides isolated from blood group H, Le^b and Le^a substances by alkaline borohydride degradation,
   L. Rovis, E. A. Kabat, M. E. A. Pereira and T. Feizi, Biochemistry $\underline{12}$, 5355-5360 (1973).
   Two recent papers which serve as examples of the outstanding and extensive studies of Kabat and his coworkers on the chemical structure of blood group substances.

13

BLOOD GROUP SUBSTANCES - II:

GENETIC CONTROL AND BIOSYNTHESIS

Family studies over the past 60 years have shown that the inheritance of ABO(H) and Lewis blood types, and hence the ability to form the structures shown in Table 14, is controlled by the action of genes at four independent loci: the *ABO*, the *LeLe*, the *Hh*, and the *Sese* loci. The small proportion of cases (less than 1%) which are inconsistent with this conclusion can be reasonably ascribed to illegitimacy, technical errors in typing or possibly mutations. When, in the late 1950's, the chemical relationship between the structures responsible for the different blood types was emerging, W. M. Watkins and W. T. J. Morgan proposed that these genes were not responsible for the production of the entire determinant but only for the formation of enzymes that attach the outer or immunodominant sugars to a central "precursor" chain common to all phenotypes. These enzymes act sequentially, the product of one being the substrate for the next, and as each sugar is added a new specificity emerges while the underlying specificity is suppressed. A similar scheme based on serological and genetic evidence obtained from family studies was proposed at the same time by R. Ceppellini. Recent work has provided abundant proof for the correctness of these proposals; the presence or absence of four specific glycosyltransferases in individuals belonging to different blood groups agrees fully with their predicted occurrence based on genotype (1,2).

The *ABO* locus is responsible for two enzymes: an N-acetylgalactosaminyltransferase specified by the *A* gene and a galactosyltransferase specified by the *B* gene. Both enzymes transfer the appropriate hexose to galactose (substituted on its C-2 by L-fucose) and form α-(1→3) linkages. The third gene at the *ABO* locus, the *O* gene, is inactive and does not produce a functional glycosyltransferase. The *Lele* locus produces one enzyme, a fucosyltransferase specified by the *Le* gene that transfers L-fucose to N-acetylglucosamine and forms α-(1→4) linkages. The *le* gene is inactive. The

Hh locus is responsible for a second fucosyltransferase that transfers L-fucose to galactose to form α-(1→2) linkages. The H gene specifies this fucosyltransferase, while its allele, the h gene, is inactive.

The $Sese$ locus does not specify an enzyme but in an unknown way is necessary for the expression of the H gene (in this sense, the formation of the fucosyltransferase specified by the H gene) in certain organs but not in others. The presence or absence of the fucosyltransferase specified by the H gene in secretory organs is the basis for their ability to secrete glycoproteins that exhibit A, B, or H specificities. As shown in Table 14, the determinants of all three specificities contain L-Fuc-α-(1→2)-Gal groupings. The 80% of the population whose secretions contain glycoproteins with A, B, or H activity possess an Se gene and are known as "secretors". The remaining 20%, the "non-secretors", have the genotype $sese$; their secreted glycoproteins lack A, B, or H specificity. The Se gene does not control the formation of the fucosyltransferase specified by the H gene in the tissue that forms the red blood cells, and its absence does not affect the formation of A, B, or H determinants of erythrocytes. The Se gene does, however, affect Lewis blood groups. Unlike the A, B, or H determinants, the determinants of the Lewis system on red cells are not integral parts of the erythrocyte membrane but, as mentioned in the last lecture, they are glycosphingolipids acquired from the serum where they circulate associated with serum lipoproteins. The origin of serum glycosphingolipids is unknown, but clearly the organ in which they are produced is affected by the *Se* gene because of the association between the secretor status and Lewis blood type: all individuals of the Le^b blood group are secretors, while nonsecretors never belong to the Le^b group. As shown in Table 14, the fucosyltransferase specified by the *H* gene is required for the synthesis of the Le^b determinant.

As pointed out earlier, it is still unsettled whether the A, B and H determinants on erythrocytes are glycolipids, glycoproteins or both. Although glycolipids with A, B, and H activity have been isolated from erythrocytes, their levels appear too low to account for all the antigenic sites. It is possible that membrane-bound glycoproteins are mainly responsible for the A, B, and H reactivity of erythrocytes. However, uncertainty as to the nature of the erythrocyte determinants is not important for the understanding of the genetic control of their biosynthesis: the glycosyltransferases specified by the genes that determine the blood types are involved in the synthesis of the carbohydrate chains of both glycolipids and glycoproteins. For example, the glycosyltransferase specified by the A gene forms the GalNAc-α-(1→3)-Gal sequence shown in Table 14. In people who have blood type A, this sequence occurs in glycolipids, glycoproteins, and also in the free oligosaccharides of milk and urine. The same sequence is not found in similar material from people with blood type B or O. Since blood type A is determined by one gene, it follows that the same glycosyltransferase, the one specified by the A gene, synthesizes GalNAc-α-(1→3)-Gal structures in both glycolipids and glycoproteins as well as in free oligosaccharides.

$$\text{GalNAc-}\alpha\text{-(1→3)-Gal} \atop {2 \atop \uparrow} \atop \underline{L}\text{-Fuc-}\alpha\text{-1}$$

GLYCOSYLTRANSFERASES SPECIFIED BY THE BLOOD GROUP GENES

Final proof for the postulated scheme of the biosynthesis of the blood group determinants and its genetic control was obtained when V. Ginsburg at the NIH and W. M. Watkins in London isolated the presumed glycosyltransferases and studied their properties and distribution. These findings have also provided striking evidence for the correctness of the "one gene - one enzyme" hypothesis proposed by G. W. Beadle and

E. L. Tatum in the early 1940's.

In order to detect specific glycosyltransferases, suitable oligosaccharides should be used. Here again milk oligosaccharides proved of primary importance since their structure corresponds to those of incomplete blood group determinants (see Tables 13 and 14). Moreover, the reaction products obtained by enzymic transfer of sugar residues to milk oligosaccharides are simple and can be characterized chemically. Using milk oligosaccharides as acceptors, and different sugar nucleotides as donors, glycosyltransferases specific for the synthesis of A, B, H, and Le substances were found in human milk, in serum, gastric mucosal linings and in submaxillary glands. More recently, glycoproteins (e.g. certain pig submaxillar mucins) have been shown to serve as acceptors for the attachment of blood type specific sugars by these enzymes. Of special significance is the finding that the glycosyltransferases can convert blood group O erythrocytes into A or B type, when the suitable enzyme and glycosyl donor are used.

THE *H* GENE ENZYME. The *H* gene specifies a fucosyltransferase that catalyzes the following reaction:

$$\text{GDP-}\underline{\text{L}}\text{-Fuc} + \text{Gal-}\beta\text{-}(1\to) \xrightarrow{H \text{ enzyme}} \underline{\text{L}}\text{-Fuc-}\alpha\text{-}(1\to 2)\text{-}\beta\text{-}(1\to) + \text{GDP}$$

The enzyme has been detected in milk and submaxillary gland preparations from secretors but not in similar preparations from non-secretors, as both the mammary and submaxillary glands are under the control of the *Se* gene. The presence or absence of this enzyme in secretory organs is the biochemical basis for secretor status. Most individuals have an *H* gene, whereas the few who do not (the *hh* genotype) belong to the "Bombay" phenotype; they are presumably unable to form the *H* enzyme in any tissue and are unable to synthesize any of the structures in Table 14, except the Le^a-active structure, and then only if they posses the *Le* gene. Their erythrocytes are devoid of H-active structures.

THE *Le* GENE ENZYME. The *Le* gene specifies a fucosyltransferase that catalyzes the following reaction:

$$\text{GDP-}\underline{L}\text{-Fuc} + \ldots\text{GlcNAc-}\beta\text{-}(1\rightarrow) \xrightarrow{\text{Lewis enzyme}} \ldots\text{GlcNAc-}\beta\text{-}(1\rightarrow) + \text{GDP}$$
$$\underset{\underline{L}\text{-Fuc-}\alpha\text{-1}}{\overset{4}{\uparrow}}$$

The enzyme has been found in donors belonging to blood group Le^a or Le^b, but not in donors who belong to neither group, the "Lewis negative" phenotype. Since both Le^a and Le^b determinants contain $\underline{L}$-Fuc-α-(1→4)-GlcNAc groupings (see Table 14), it follows that individuals who are "Lewis negative" (about 7% of the population) would be unable to make either structure.

This glycosyltransferase can also transfer $\underline{L}$-fucose to a glycoprotein without blood group activity to yield a product precipitable with anti-Le^a sera.

THE A GENE ENZYME. The A gene specifies an N-acetylgalactosaminyltransferase that catalyzes the following reaction:

$$\text{UDP-GalNAc} + \text{Gal-}\beta\text{-}(1\rightarrow) \xrightarrow{A \text{ enzyme}} \text{GalNAc-}\alpha\text{-}(1\rightarrow 3)\text{-Gal-}\beta\text{-}(1\rightarrow) + \text{UDP}$$
$$\underset{\underline{L}\text{-Fuc-}\alpha\text{-}1}{\overset{2}{\uparrow}} \qquad\qquad \underset{\underline{L}\text{-Fuc-}\alpha\text{-}1}{\overset{2}{\uparrow}}$$

The enzyme is present in donors with an A or AB blood type and is absent from donors with B or O blood type. It has a strict acceptor requirement in that it will transfer N-acetylgalactosamine to galactose only if the galactose is substituted on the 2 position with $\underline{L}$-fucose. This stringent specificity explains why the Se gene not only controls the formation of H-active glycoproteins but also of A-active (and B-active) glycoproteins as well. Individuals with blood type A who do not have the Se gene have normal levels of the A enzyme. Nevertheless, they are not able to synthesize A-active soluble glycoproteins because the necessary $\underline{L}$-Fuc-α-(1→2)-Gal grouping required by the A enzyme for the acceptor is missing from their secreted glycoproteins. Similarly, the absence of an H gene can modify the expression of the A or B gene: individuals belonging to the "Bombay" blood group cannot make A or B determinants on their red cells, even though they may possess the *A* or *B* gene, and thus also the corresponding glycosyltransferases.

Of special interest is the inability of lacto-difucotetraose and lacto-*N*-difucohexaose I to accept *N*-acetylgalactosamine when incubated with the *A* gene enzyme, even though their penultimate galactosyl residues are substituted on the 2 position with L-fucose. Presumably, the second L-fucose residue on the adjacent sugar sterically hinders the enzyme. Lacto-difucotetraose and lacto-*N*-difucohexaose I resemble the Le^b determinant (see Tables 13,14) and are potent hapten inhibitors of the agglutination of Le^b erythrocytes by anti-Le^b serum. It is probable that once Le^b-active structures are formed *in vivo*, they will persist as they cannot act as acceptors for further addition of sugars. If this is true, the formation of the structure proposed for the A determinant,

$$\text{GalNAc-}\alpha\text{-}(1\rightarrow 3)\text{-Gal-}\beta\text{-}(1\rightarrow 3)\text{-GlcNAc-}\beta\text{-}(1\rightarrow)$$
$$\underset{\underset{\text{L-Fuc-}\alpha\text{-1}}{\uparrow}}{2} \quad \underset{\underset{\text{L-Fuc-}\alpha\text{-1}}{\uparrow}}{4}$$

would involve a strict sequence of addition of the three terminal sugars as dictated by enzyme specificities. First, L-fucose would be added to the galactose; then *N*-acetylgalactosamine would be added to the same galactose; and finally, the second L-fucose would be added to N-acetylglucosamine.

The difference between blood group A_1 and A_2 is not yet clear. Earlier no difference was found between the A_1 or A_2 enzymes *in vitro*, including their ability to add N-acetylgalactosamine to oligosaccharides with the following structures:

Type 1
$$\text{Gal-}\beta\text{-}(1\rightarrow3)\text{-GlcNAc}\underset{\underset{\text{L-Fuc-}\alpha\text{-1}}{\uparrow}}{2}\ldots$$

and Type 2
$$\text{Gal-}\beta\text{-}(1\rightarrow4)\text{-GlcNAc}\underset{\underset{\text{L-Fuc-}\alpha\text{-1}}{\uparrow}}{2}\ldots$$

Recent detailed kinetic studies by H. Schachter and his coworkers (A1) of the N-acetylgalactosaminyltransferases in human serum, revealed a lower activity of the transferase in serum from A_2 donors than in serum from A_1 donors. Also, the Michaelis constants, cation requirements and pH optima of the A_1 and A_2-gene-specified enzymes, with 2'-fucosyl-lactose (containing the sequence Gal-β-(1→4)-Glc) and lacto-N-fucopentaose I (with the Gal-β-(1→3)-GlcNAc sequence) as substrates, were different. No absolute difference in specificity between the two substrates was, however, observed.

These results are compatible with the view that the

A_2 gene produces a mutant enzyme that carries out the same function as the A_1-gene-specified transferase, but does so less effectively so that fewer of the precursor H structures are changed to A active determinants. That is, the difference in the final product is essentially quantitative and not qualitative.

THE B GENE ENZYME. The B gene specifies a galactosyltransferase that catalyzes a reaction identical to the one carried out by the enzyme specified by the A gene, except that galactose is transferred in place of N-acetylgalactosamine, as follows:

$$\text{UDP-Gal} + \underset{\underset{\underline{\text{L}}\text{-Fuc-}\alpha\text{-1}}{\uparrow 2}}{\text{Gal-}\beta\text{-(1}\rightarrow\text{)}} \xrightarrow{B \text{ enzyme}} \underset{\underset{\underline{\text{L}}\text{-Fuc-}\alpha\text{-1}}{\uparrow 2}}{\text{Gal-}\alpha\text{-(1}\rightarrow\text{3)-Gal}} \ldots + \text{UDP}$$

It is present in secretions from donors of B or AB blood type and is absent from the milk of donors with A or O blood type. Its occurrence, like that of the A gene enzyme, is independent of secretor status. Like the A gene enzyme, the B gene enzyme requires the galactose acceptor to be substituted with $\underline{\text{L}}$-fucose on the 2 position. The acceptor specificity for other milk oligosaccharides of the two enzymes is the same; and modification of the expression of the B gene by the Se gene or the H gene can be explained in the same way.

As the *A* gene and the *B* gene are alleles, the primary structure of the two enzymes may be quite similar. The third allele at the *ABO* locus, the *O* gene, is inactive, presumably producing a non-functional A or B "enzyme".

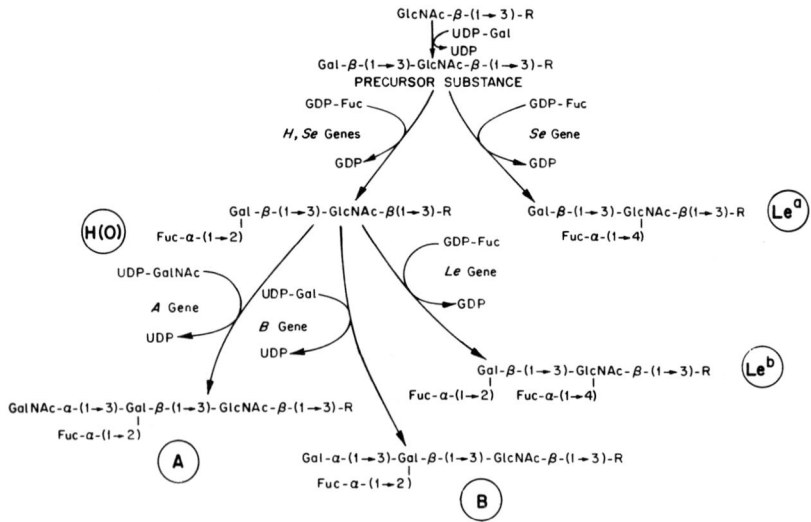

Figure 46 Genetic control of the synthesis of the A,B,H and Lewis (Le^a and Le^b) blood group determinants of Type 1.

The action of the four enzymes described above is summarized in Fig. 46. The carbohydrate structures that they form are depicted, along with their associated serologic activity. Clearly, these determinants of blood type are secondary gene products, in that the primary gene products are the enzymes and it is these enzymes, working in concert, that determine which structures are formed. This mechanism of synthesis has an important biological consequence: it provides a biochemical explanation for antigens produced by "gene interaction" - that is to say, antigens present in a hybrid that are not found in either parent. The Le^b antigen is one of these. From family studies, Ceppellini proposed that the Le^b antigen was an interaction product of two genes which, according to Watkins and Morgan, are the *Le* and the *H*, as well as the *Se* genes (the latter, you will recall, controls the expression of the *H* gene). As shown in Fig. 46, both genes are necessary to produce the two fucosyltransferases required for the synthesis of the Le^b antigen. A child who inherited *Le* and *H* genes from one parent who happened to have the genotype *sese*, and an *Se* gene from the other parent who happened to have the genotype *lele*, would synthesize a structure that neither parent could make alone. The whole system resembles the complex interaction of genes that determine the numerous specificities of the cell wall antigens (*O*-specific polysaccharides) of the *Salmonella*, to be discussed later on in this course. There are other reports of interaction of genes resulting in new cell antigens.

The formation of unique carbohydrate structures by

the concerted action of glycosyltransferases provided by both parents may be a factor in the expression of individuality. These structures would reflect not only the various glycosyltransferases that are present but, because the reactions are competitive, they would also reflect the relative activities of these enzymes as well. For example, as mentioned above, the *A* gene enzyme and the *B* gene enzyme probably cannot use as acceptors those carbohydrate chains that contain $\underline{L}$-Fuc-α-(1→4)-GlcNAc groupings. If the level of the *Le* gene enzyme that makes this grouping is high relative to the *H* gene enzyme, it is clear from Fig. 46 that more Le^a and Le^b determinants would be formed than A, B, or H determinants. If the *Le* gene enzyme is less active than the *H* gene enzyme, the reverse would be true. Coupled with the enzyme specificity of subsequent steps, competition of this type would tend to generate heterogeneity among the carbohydrate chains of glycoproteins and, along with the incomplete synthesis of these chains, would contribute to the multiple blood group specificities that are exhibited by single glycoprotein molecules.

BLOOD GROUP M AND N SPECIFICITIES AND THE MAJOR GLYCOPROTEIN OF THE HUMAN ERYTHROCYTE MEMBRANE

The second human blood group system, the MN system,

was discovered by Landsteiner and Levine in 1927. The M and N antigens are found only on the erythrocyte membrane, and not in soluble form, in serum or in secretions. They were first detected by antisera obtained by immunization of rabbits with human red blood cells, and subsequently by plant lectins, such as that from *Vicia graminea* which possesses anti-N activity.

The first insight into the chemical nature of the M and N specificities came from the observation made by G. Springer in 1958 that neuraminidases from influenza virus or from *Vibrio cholerae* abolished the M and N activities of human erythrocytes in addition to destroying their influenza virus receptor activity. It was concluded that sialyl residues serve not only as receptor sites for the influenza virus hemagglutinin, but are also essential for the M and N activities.

Incubation of erythrocytes with proteolytic enzymes, such as trypsin or pronase, releases most of their carbohydrate in the form of a small number of glycopeptides (molecular weights of up to 10,000) without impairing the integrity of the membrane. Among the released glycopeptides are the M and N antigens. Much of our knowledge of the M and N determinants is based on studies of such tryptic fragments, as well as of the major glycoprotein of the human erythrocyte membrane, also known as glycophorin (3). These studies, carried out during the last decade in a number of laboratories in Poland, Germany and the U.S., have established that the M and N receptors are oligosaccharide constitutents of glycophorin. This glycoprotein also carries the receptors for influenza virus, as well as for a number of lectins. I shall therefore leave the M and N substances for a short while, to summarize our present views of this important glycoprotein and its place in the membrane.

The human erythrocyte membrane (4,5) consists by weight of 52% protein, 40% lipid and 8% carbohydrate. It contains at least half a dozen glycoproteins, which carry over 90% of its carbohydrate, the rest being in the form of glycosphingolipids. The glycoproteins can be solubilized by the use of aqueous phenol, aqueous pyridine, sodium dodecylsulfate or lithium diiodosalicylate. The major glycoprotein, glycophorin, comprises about 5% of the total protein of the erythrocyte membrane, and contains about 80% of the membrane sialic acid. It was isolated from membrane extracts first by conventional methods, and very recently also by affinity chromatography on a column of wheat germ agglutinin bound to Sepharose (A2). Purified glycophorin consists, by weight, of about 40% protein and 60% carbohydrate (including 25% N-acetylneuraminic acid). Glycophorin readily undergoes aggregation and is usually isolated in a highly aggregated form with an apparent molecular weight of $\sim$ 400,000. Because of this, and the fact that estimation of the molecular size of highly glycosylated proteins is subject to many errors (in particular when polyacrylamide gel electrophoresis is used for this purpose; see pp.51-52), different values for the molecular weight of glycophorin appeared in the literature. It now appears certain that in its disaggregated form, glycophorin has a molecular weight of 29,000 (A2), in excellent agreement with the value of 31,000 reported by R. H. Kathan and R. J. Winzler in 1963. On this basis, the protein moiety of glycophorin consists of 131 amino acids, and it carries about 100 sugar residues. Each human erythrocyte thus contains about 500,000 molecules of this glycoprotein. The complete amino acid sequence and the oligosaccharide attachment sites of glycophorin have recently been determined in the laboratory of V. T. Marchesi at Yale University.

The carbohydrate of glycophorin is arranged in 16 oligosaccharide chains that are attached to the polypeptide backbone either by O-glycosidic linkages to serine or threonine residues (15 chains), or by N-glycosidic linkages to asparagine residues (one chain). The O-glycosidically linked units were previously isolated as reduced oligosaccharides after alkaline borohydride treatment of the glycoprotein, or of M and N active glycopeptides obtained from trypsin digests of the erythrocyte membrane. The largest oligosaccharide thus released was a tetrasaccharide (Fig. 47[1]). Upon pronase digestion of the glycopeptides remaining after the alkaline borohydride treatment, a number of low molecular weight glycopeptides was isolated. In one of these the branched carbohydrate structure given in Fig. 47[2] was found.

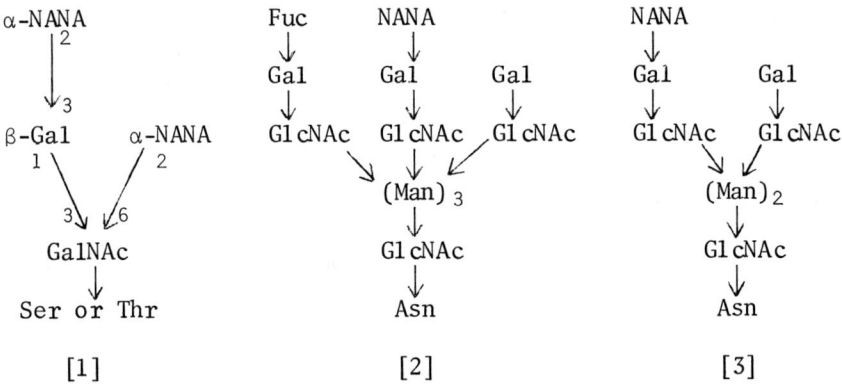

Figure 47 Proposed structures of oligosaccharide constituents of human erythrocyte membranes. [1] Alkali labile and [2] alkali stable oligosaccharides isolated from glycophorin or from tryptic digests of membranes. [3] Structure of PHA receptor from human erythrocyte membrane.

All the carbohydrate units are linked to the N-terminal half (approximately 64 residues) of the single polypeptide chain of glycophorin, and are exposed to the external environment of the cell. Their sialic acid residues contribute to the high negative charge of the erythrocyte surface, the main function of which appears to keep the erythrocytes apart from one another. This arrangement makes the carbohydrate chain available to the interaction with anti-M and anti-N reagents, with influenza virus and with lectins. It also explains why trypsin releases from erythrocytes glycopeptides that bind to these agents. The C-terminal end (approximately 35 amino acid residues) of the polypeptide appears to be located internal to the lipid barrier of the membrane, and possibly extends into the cytoplasm of the cell. It has an unusually high content of hydrophilic amino acids. The segment connecting these two portions is composed of 32 residues, almost all of which are of nonpolar and hydrophobic amino acids; apparently, this segment spans the lipid region of the membrane. It is believed that other glycoproteins are also arranged in the membrane in a similar manner, extending asymmetrically across the membrane, with their carbohydrate moieties exposed to the external environment. Models of glycophorin, with the M and N antigens, the receptors for the influenza virus and for several lectins

(PHA, WGA and the lectin of the mushroom *Agaricus bisporus*) have been proposed by various authors. One such model is depicted below (Fig.48). It has been claimed that glycophorin also carries the ABH blood type determinants, but this is very much in doubt.

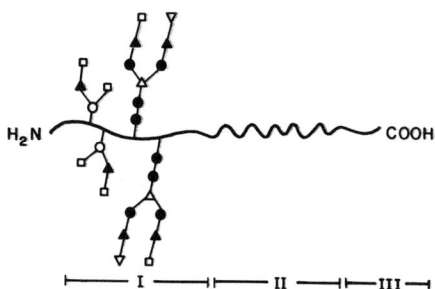

Figure 48 Schematic representation of glycophorin. I: N-Terminal segment (receptor region or external domain); II: hydrophobic region (intramembranous); III: C-terminal segment (internal domain). The branched structures depict oligosaccharide side chains.

While M and N blood group activity has clearly been attributed to glycophorin, the complete structure of these antigenic determinants has not yet been fully established. The overall composition of M, N and MN active glycoproteins or of the carbohydrate units derived from them (either alkali labile or asparagine-linked) is very similar.

Identification of the immunodeterminant structure of these blood group substances has also been hampered by the fact that the isolated oligosaccharides lack M or N activity, and the glycopeptides show only low activity. This strongly suggests that the peptide portion of the glycoprotein plays a major role in the interaction of the M and N blood group substances with the corresponding antibodies. Indeed, blocking of the amino groups of the erythrocyte glycoprotein results in complete loss of its M and N specificity. It is also well established that the M and N activities of glycophorin or of isolated glycopeptides, are highly dependent on their state of aggregation, both activities being the highest for the largest molecules or molecular aggregates. This is also true for the influenza virus inhibitory activity of the same substances.

Based on enzymic degradation studies and serological experiments, Springer has recently put forward a structure for the human group M and N antigens (A4). According to his model, the M determinant contains two sialic acid residues, both linked to β-galactosyl units; in the N determinant, one of the sialic acid residues is missing, and the specificity is determined by the unmasked β-galactosyl residue together with the remaining terminal sialyl residue:

$$\begin{array}{cc}
\text{NANA} \xrightarrow{\alpha} \text{Gal} \searrow^{\beta} & \text{NANA} \xrightarrow{\alpha} \text{Gal} \searrow^{\beta} \\
\quad\quad\quad\quad\quad\quad \text{GalNAc} & \quad\quad\quad\quad\quad\quad \text{GalNAc} \\
\text{NANA} \xrightarrow{\alpha} \text{Gal} \nearrow^{\beta} & \quad\quad\quad\quad \text{Gal} \nearrow^{\beta} \\
\text{M-specific} & \text{N-specific}
\end{array}$$

Please note that the sequence NANA-Gal-GalNAc occurs in the glycopeptides isolated from glycophorin (Fig.47[1]). The proposed structure is also in agreement with other data on the M and N specificities. Thus, mild acid treatment of M substance converts it readily to N substance, with removal of only sialic acid. Uncovering of the N-specific structure parallels the appearance of terminal β-galactosyl residues; the removal of the latter results in destruction of N-specificity.

According to Springer, blood group N is the immediate precursor of blood group M. Thus these two antigenic specificities are not determined by two allelic genes as was hitherto believed. Rather, the relationship between the *M* and *N* genes is like that between the *AB* and *H* genes, which I have discussed at length earlier.

REFERENCES

Review

1. *See references 2, 3 and 4 to lecture 12 - in particular the review by V. Ginsburg (ref.2) on which much of the first part of lecture 13 is based.*

2. The biosynthesis of animal glycoproteins, H. Schachter and L. Rodén, in Metabolic Conjugation and Metabolic Hydrolysis (Ed. W. H. Fishman), Vol.3, Academic Press, 1973, pp.1-149.
    pp.88-101 of this excellent review deal with biosynthesis of blood group substances, including the

transfer of immunodeterminant sugar residues to high molecular weight acceptors.

3. Molecular features of the major glycoprotein of the human erythrocyte membrane,
V. T. Marchesi, R. L. Jackson, J. P. Segrest and I. Kahane, Fed. Proc. 32, 1833-1837 (1973).

4. The organization of proteins in the human red blood cell membrane,
T. L. Steck, J. Cell. Biol. 62, 1-19 (1974).
Up-to-date review, highly recommended.

5. Membrane structure: some general principles,
M. S. Bretscher, Science 181, 622-629 (1973).
Critical and provocative; contains concise and interesting section on erythrocyte membrane.

Specific articles

A1. Qualitative differences in the N-acetyl-$\underline{D}$-galactosaminyltransferases produced by human A^1 and A^2 genes,
H. Schachter, M. A. Michaels, C. A. Tilley, M. C. Crookston and J. H. Crookston, Proc. Natl. Acad. Sci. USA 70, 220-224 (1973).

A2. Isolation of the receptors for wheat germ agglutinin and the *Ricinus communis* lectins from human erythrocytes using affinity chromatography,
W. L. Adair and S. Kornfeld, J. Biol. Chem. 249, 4696-4704 (1974).
The wheat germ agglutinin receptor is located on the carbohydrate side chain of glycophorin.

A3. The molecular weight of the major glycoprotein from the human erythrocyte membrane,
S. P. Grefrath and J. A. Reynolds, Proc. Natl. Acad. Sci. USA 71, 3913-3916 (1974).

A4. Common precursors of human blood group MN specificities,
G. F. Springer and P. R. Desai, Biochem. Biophys. Res. Commun. 61, 470-475 (1974).

14

MUCOPOLYSACCHARIDES (PROTEOGLYCANS)

- I: CHEMICAL STRUCTURE

The glycoproteins that I have discussed till now are all similar in that their oligosaccharide chains are made up of no more than 15-20 monosaccharide units, and they do not contain any repeating oligosaccharide structures. Today I shall deal with a special class of glycoproteins, the mucopolysaccharides or proteoglycans (1-6), that consist typically of a protein core to which are covalently linked many long chain linear heteropolysaccharides, made up largely of disaccharide repeating units. In the disaccharide one sugar is always a hexosamine, either glucosamine or galactosamine, most commonly in its N-acetylated form; and the other a non-nitrogeneous sugar, glucuronic acid or $\underline{L}$-iduronic acid (p.44). Another constituent common to most of the mucopolysaccharides are sulfate groups, linked by ester bonds to the hydroxyl groups of their monosaccharide constituents, and in some cases also by amide linkages to the amino groups of glucosamine. By virtue of their carboxyl and sulfate groups, mucopolysaccharides are highly charged polyanions. As we shall see, polymers that do not conform in all respects to the above structural definitions, are also considered as mucopolysaccharides.

The name mucopolysaccharide, or acid mucopolysaccharide, was coined in the 1930's by Karl Meyer from Columbia University who did most of the pioneering work in this field. It took a long time to realise that mucopolysaccharides are indeed glycoproteins, since the early methods of preparation yielded products which contained little or no protein. To distinguish them from other glycoproteins, they are referred to as proteoglycans and their carbohydrate chains as glycosaminoglycans. However these terms have not yet gained wide acceptance and the name mucopolysaccharide is still much in use.

Mucopolysaccharides occur in many animal tissues and fluids. A rich source of mucopolysaccharides is connective tissue, such as skin, bone, cartilage and ligaments where, together with collagen or elastin fibers, they form the matrix or "ground substance" in which the connective tissue cells (or fibroblasts) are embedded. Relatively high proportions of these glycoproteins are also found in arterial walls, the umbilical cord, the vitreous humor in the globe of the eye and in synovial fluids. In some cases mucopolysaccharides are present inside cells, primarily in mast cells (a special type of connective tissue cell), and in circulating cells such as granulocytes and platelets. Recently they have been detected in cell membranes as well. The pattern of distribution of mucopolysaccharides in the same tissue or fluid changes with maturation and aging. Disturbances in the normal distribution of mucopolysaccharides lead to serious clinical abnormalities characterizing various diseases of mucopolysaccharide metabolism, which I shall discuss in a later lecture. Abnormalities in the level and nature of mucopolysaccharides occur in many other conditions, including skin diseases, rheumatoid arthritis and ocular diseases.

The biological functions of mucopolysaccharides are not known with certainty. They are presumed to act in stabilizing and supporting the fibrous and cellular elements of tissues, to contribute to the load-bearing characteristics of anatomical surfaces, and to serve as lubricants. The ability of mucopolysaccharides to fulfil these functions is a consequence of their macromolecular structure and electrical charge. These materials form solutions with high viscosity and elasticity, in which each molecule occupies a volume that is 1,000 to 10,000 times larger than that of the total volume of its sugar units. Such molecules tend to form spheres, more or less filled with water, which, because of their carboxyl and sulfate groups, may behave as ion-exchange resins and serve as traps for cations such as calcium. In this way, they are believed to contribute to the maintenance of water and salt balance in the body.

In the past, mucopolysaccharides were commonly extracted from tissues after proteolytic digestion, which led to their isolation in the form of polysaccharides linked to small peptides, or by treatment with alkali. The latter method is now rarely used, as it leads to cleavage of the carbohydrate-peptide bond in most of these glycoproteins.

For studies of the intact molecules, it is important to extract the tissues under the mildest conditions possible. Although salt solutions are most often used for this purpose, in some cases extraction can be achieved only by using concentrated urea, guanidinium chloride, or other denaturing agents. The ease of extraction and subsequent fractionation varies considerably with the tissue investigated. Fractionation of the extract is carried out by ion-exchange chromatography, or by density gradient centrifugation. The latter technique is applicable for this purpose since glycoprotein molecules are more dense than those of non-glycosylated proteins (p. 49). Mucopolysaccharides can also be fractionated according to their polyanionic character (charge density and molecular weight) with the aid of quaternary ammonium salts such as cetylpyridinium chloride. This approach, developed in the 1950's, proved extremely successful and is still in use.

The classification of mucopolysaccharides is based mainly on the work of Karl Meyer and his associates. Currently we distinguish at least six mucopolysaccharides: hyaluronic acid, chondroitin 4- and 6-sulfate (previously known as chondroitin sulfates A and C, respectively), dermatan sulfate (chondroitin sulfate B), heparin, heparan sulfate (heparitin sulfate) and keratan sulfate (or keratosulfate). Table 15 gives the composition of the repeating units of the polysaccharides which have been recognized so far. In addition to their major sugar constituents, the two hexosamines and the two uronic acids, mucopolysaccharides have as integral components other monosaccharides, including sialic acid, mannose, fucose, galactose and xylose. With the exception of galactose, however, these sugars are not part of the characteristic repeating disaccharides and occur either as

Table 15

Major components of the polysaccharide portion of mucopolysaccharides (proteoglycans)

	Usual molecular weight of polysaccharide chain	Component sugars	Location of sulfate	Linkage	Major Source
Hyaluronic acid[a]	$1 - 3 \times 10^6$	N-acetylglucosamine glucuronic acid	—	$\beta-(1\rightarrow 4)$ $\beta-(1\rightarrow 3)$	synovial fluid, vitreous humor of the eye, umbilical cord, cock's comb
Chondroitin 4-sulfate (chondroitin sulfate A)	$2 - 5 \times 10^4$	N-acetylgalactosamine glucuronic acid	4	$\beta-(1\rightarrow 4)$ $\beta-(1\rightarrow 3)$	human cartilage, aorta
Chondroitin 6-sulfate (chondroitin sulfate C)	$2 - 5 \times 10^4$	N-acetylgalactosamine glucuronic acid	6	$\beta-(1\rightarrow 4)$ $\beta-(1\rightarrow 3)$	heart valves
Dermatan sulfate (chondroitin sulfate B)	$2 - 5 \times 10^4$	N-acetylgalactosamine iduronic acid glucuronic acid	4	$\beta-(1\rightarrow 4)$ $\alpha-(1\rightarrow 3)$[b] $\beta-(1\rightarrow 3)$	skin, blood vessels, heart valves
Heparin	$1 - 3 \times 10^4$	glucosamine glucuronic acid iduronic acid	3,6,N 2	$\alpha-(1\rightarrow 4)$ $\beta-(1\rightarrow 4)$ $\alpha-(1\rightarrow 4)$[b]	lung, mast cells
Heparan sulfate (heparitin sulfate)	$2 - 10 \times 10^3$	glucosamine N-acetylglucosamine glucuronic acid iduronic acid	N ? 3,6 2	$\alpha-(1\rightarrow 4)$ $\beta-(1\rightarrow 4)$ $\alpha-(1\rightarrow 4)$[b]	blood vessels, cell surfaces
Keratan sulfate	$5 - 20 \times 10^3$	N-acetylglucosamine galactose	6 6	$\beta-(1\rightarrow 3)$ $\beta-(1\rightarrow 4)$	cornea of the eye, nucleus pulposus, cartilage

[a] The attachment of hyaluronic acid to protein has not been demonstrated unequivocally.

[b] This linkage of L-iduronic acid, identical to the β-linkage of D-glucuronic acid. However, iduronic acid is of the L rather than D configuration, which results in this bond being designated as α rather than β.

side branches or as constituents of the specific carbohydrate-protein linkage region, which I shall discuss in more detail somewhat later. Galactose is part of the repeating unit in keratan sulfate, where it is linked to N-acetylglucosamine, and thus occupies the position normally held by a uronic acid. Keratan sulfate is the only compound classified as a mucopolysaccharide or proteoglycan which is devoid of uronic acid in its main chain. It may also contain a branched heterosaccharide consisting of sialic acid, fucose, mannose and N-acetylgalactosamine. In this, as well as some other respects, keratan sulfate is more akin to the typical glycoproteins.

Two classical methods of structure determination were responsible for most of our early knowledge of the chemical structure of glycosaminoglycans. The oldest method is based on the degradation of the polysaccharide into oligosaccharides using acids or enzymes. By partial acid hydrolysis the disaccharide chondrosine was isolated from chondroitin sulfate as early as 1914. However, the difficulty in elucidating the structure of chondrosine was so great that only in 1955 was it definitely characterized as glucuronyl-β-(1→3)-galactosamine. A similar disaccharide, glucuronyl-β-(1→3)-glucosamine (hyalobiuronic acid) was isolated in

1951 from a partial hydrolysate of hyaluronic acid. One of the disadvantages of acid degradation is that it rapidly removes the sulfate groups making the determination of their location impossible.

Application of the second method, the methylation procedure, to mucopolysaccharides, met with success only during the 1950's, after all methylated derivatives of glucosamine and galactosamine, which were required as reference compounds, became available. Most of these compounds have been synthesized by Roger Jeanloz and his group at the Massachusetts General Hospital in Boston. An important advantage of methylation is that it does not cause cleavage or migration of the sulfate groups. Methylation studies of the intact mucopolysaccharides and their desulfated derivatives have given conclusive evidence on the location of the sulfate groups and on the sites of glycosidic substitution.

Recent work, using improved methylation techniques coupled with gas liquid chromatography, nuclear magnetic resonance, degradation by different enzymes, etc., has led to a more complete characterization of the structure of mucopolysaccharides, and to the recognition that in many cases deviations occur from the idealized version of the simple repeating disaccharide pattern. I shall now review in more detail the characteristics of some mucopolysaccharides, as a basis for our discussion of their biosynthesis and metabolism under normal and pathological conditions.

HYALURONIC ACID

Hyaluronic acid, isolated for the first time in a pure state by Karl Meyer in 1934 from bovine vitreous humor, possesses the least complex chemical structure of connective tissue mucopolysaccharides. It can be isolated intact from most tissues and fluids in high yield without proteolysis or alkali treatment. In fact, there is no conclusive evidence that hyaluronic acid is glycoprotein. It is nevertheless classified as a mucopolysaccharide because of its pronounced structural similarity to the other polymers of this group. I have already mentioned the isolation of hyalobiuronic acid from its acid hydrolysates. Enzymic hydrolysis of hyaluronic acid with testis hyaluronidase, gave a tetrasaccharide in high yield, composed of two units of N-acetylhyalobiuronic acid linked by a 4-O-$\underline{\text{D}}$-glucosaminyl bond. On the other hand, degradation of hyaluronic acid by bacterial hyaluronidases led to the formation of an unsaturated disaccharide, a derivative of hyalobiuronic acid with a double bond between C-4 and C-5 in the glucuronic acid residue (Fig.49).

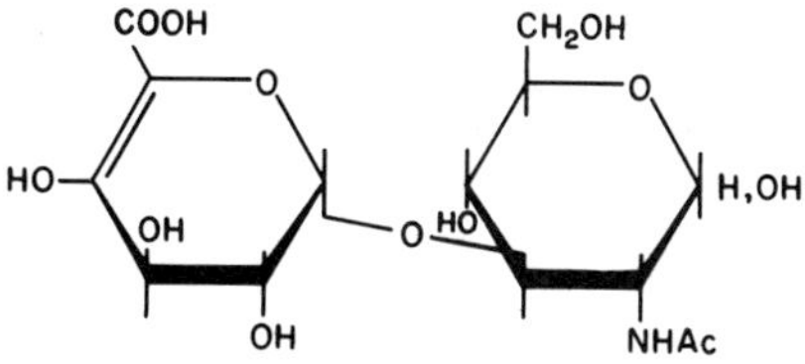

Figure 49 Δ 4,5-glucuronyl-β-(1→3)-*N*-acetylglucosamine

Thus, cleavage of hyaluronic by bacterial hyaluronidases occurs not by hydrolysis but by a process of elimination and these enzymes are therefore classified as eliminases.

It is interesting that a similar unsaturated disaccharide (in which *N*-acetylglucosamine is replaced by *N*-acetylgalactosamine) has been obtained by degradation of chondroitin 4-sulfate, chondroitin 6-sulfate and of dermatan sulfate, by bacterial enzyme preparations possessing both eliminase and sulfatase activities, thus establishing the close structural relationship between these mucopolysaccharides and hyaluronic acid.

On the basis of these and other experiments, it has been established that hyaluronic acid is made up exclusively of repeating GlcUA-β-(1→3)-GlcNAc units, linked by β-(1→4) bonds. In some of its properties, hyaluronic acid is unique among the mucopolysaccharides. As already mentioned, it is not certain at all whether it contains covalently-linked protein. It is also the only mucopolysaccharide that is not limited to animal tissues, but is produced by some strains of bacteria as well. Another property which distinguishes hyaluronic acid from the other polymers of this class is the length of its chains, which may reach a molecular weight of several millions, as compared to 10-50,000 for the carbohydrate side chains of most other mucopolysaccharides.

The polyelectrolyte character of hyaluronic acid, the large volume it occupies in solution (1 g of hyaluronic acid occupies in water a volume of about 1 litre, so that no other macromolecules can dissolve in it) and its enormous molecular length, all contribute to its being a remarkable shock absorber, as in jumping, protecting the surface layers of the joints. Recent evidence, based mainly on X-ray diffraction studies of hyaluronic acid films, indicates that this polymer chain may form a double helix structure.

CHONDROITIN SULFATES

The chondroitin sulfates are the most abundant mucopolysaccharides in the body, and occur both in skeletal and soft tissue. Much of our knowledge of the structure and metabolism of mucopolysaccharides is derived from studies of chondroitin 4-sulfate, the first sulfated compound of this class to be isolated in a pure state from cartilage.

The repeating disaccharide unit of chondroitin 4-sulfate is composed of glucuronic acid and N-acetylgalactosamine-4-sulfate and has the structure shown in Fig. 50.

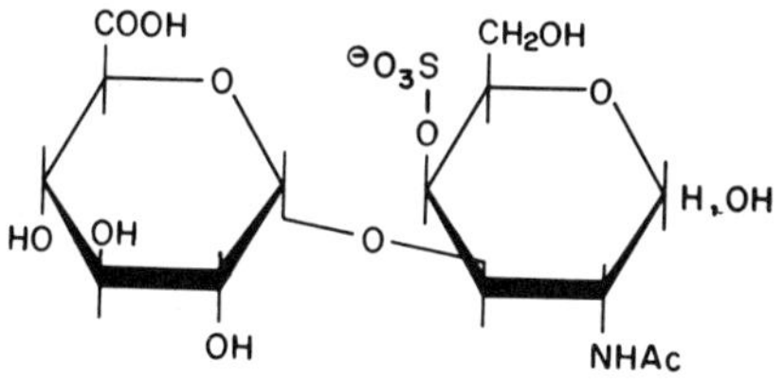

Figure 50   Structure of the repeating disaccharide unit of chondroitin 4-sulfate.

Chondroitin 6-sulfate has a carbohydrate structure identical to that of chondroitin 4-sulfate, but the sulfate group is located on C-6 of the N-acetylgalactosamine moiety. Each polysaccharide chain contains between 30 and 50 such disaccharide units, corresponding to a molecular weight of 15,000-25,000.

Sometimes deviations from the idealized structure in Fig. 50 occur, and careful analyses of products of enzymic degradation have revealed the presence of a small proportion of unsulfated disaccharide units in many chondroitin sulfate preparations. Oversulfation is also observed, with the extra sulfate either on the glucuronic acid or on the galactosamine. Another type of structural heterogeneity is the occurrence of hybrid molecules containing both 4-sulfated and 6-sulfated galactosamine residues.

THE LINKAGE REGION (5)

Neither of the two components of the repeating disaccharide mediates the linkage of the chondroitin sulfate chains to protein. Mainly due to the work of Lennart Rodén and his collaborators at the University of Chicago, the structure of the linkage region between chondroitin 4-sulfate and the protein moiety is now well characterized. Unlike the main part of the chain, which is digested by hyaluronidase, the linkage region is resistant to the action of this enzyme. It was therefore possible to isolate the linkage region by gel filtration of the degradation products obtained by hyaluronidase digestion of proteoglycans (containing chondroitin sulfate) followed by proteolysis (or alternatively, using proteolytic enzymes first, and then hyaluronidase) (Fig.51). The structure of the isolated linkage region was established with the aid of specific enzymes and comparison with derivatives of known structure (Fig. 52).

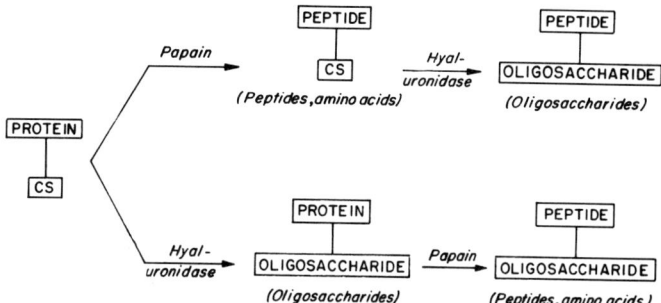

Figure 51 Alternate pathways for the enzymic degradation of chondroitin sulfate (CS). From ref.5.

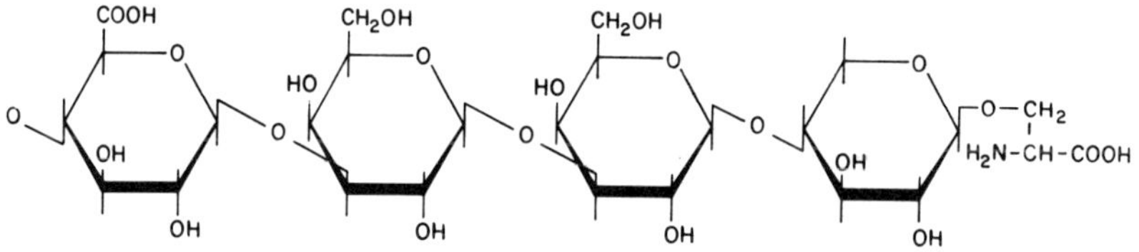

Figure 52 Carbohydrate-peptide linkage region of chondroitin 4-sulfate.

It was found that the terminal sugar, xylose, is linked to the hydroxyl of serine by an O-glycosidic linkage, which is alkali labile. The specific carbohydrate-protein linkage region also contains two galactose residues and one glucuronic acid, after which the first repeating disaccharide unit follows.

The same linkage region has also been found in

chondroitin 6-sulfate, dermatan sulfate, heparin and heparan sulfate.

KERATAN SULFATE

In proteoglycans from adult mammalian cartilage, chondroitin sulfate is always accompanied by keratan sulfate. The latter unusual proteoglycan appears to be exclusive to cartilage, invertebral discs (nucleus pulposus) (up to 10% dry weight in humans) and cornea. Keratan sulfate does not contain uronic acids, and its repeating disaccharide unit is composed of alternating galactose and N-acetylglucosamine residues linked β-$(1\rightarrow 4)$ and β-$(1\rightarrow 3)$, respectively. This repeating sequence is sulfated to various degrees at the C-6 position of both sugar residues. Other sugars that appear to be components of certain keratan sulfates are fucose, mannose, sialic acid and N-acetylgalactosamine, but their exact location in the proteoglycan molecule is unknown. The linkage to protein of corneal and of cartilage keratan sulfates is different. In corneal keratan sulfate it is stable to alkali, and appears to be a glycosylamine linkage between glucosamine and asparagine; in cartilage a majority of the linkages are the alkali labile O-glycosidic bonds between N-acetylgalactosamine and the hydroxyl of serine or of threonine (A1,A2).

STRUCTURAL HETEROGENEITY OF REPEATING DISACCHARIDE UNITS

Whereas the structures of hyaluronic acid and the chondroitin sulfates are now well established, certain other polysaccharides are more complicated and present problems, some of which are still under active investigation.

Heparin, discovered in 1916, is a particular case in point (Table 15). This mucopolysaccharide, isolated usually from lung, is probably the only polymer of this group which is not commonly present in connective tissue. Moreover, it possesses a wide variety of biological activities, of which the anticoagulant and the antilipemic effects are best known. Not long ago, heparin was regarded as a glucuronic acid-glucosamine polymer in which both the uronidic and hexosaminidic linkages were of the α configuration, and where all the amino groups were sulfated. However, our view of the structure of this polysaccharide has undergone a remarkable change over the past ten years.

In 1962, J. A. Cifonelli and A. Dorfman reported the presence of some L-iduronic acid in highly purified preparations of heparin. This report was met with a fair degree of scepticism, and most workers in the field believed that the iduronic acid was an artefact formed during the degradation of heparin. There is now convincing evidence that L-iduronic acid is a constituent of the heparin molecule. Thus, nuclear magnetic resonance studies on the intact polysaccharide have indicated that L-iduronic acid may account for as much as two-thirds of the total uronic acid. Under appropriate hydrolysis conditions more than half of the uronic acid may be liberated as free L-iduronic acid.

As mentioned above, the anomeric configuration of all glycosidic linkages in heparin had been assumed to be α. However, the optical rotation of heparin is significantly lower than would be expected for an entirely α-linked polymer, and recent work has shown that probably all of the glycuronidic linkages are actually of the β configuration. (Examination of the structure of glucuronic and L-iduronic acid clearly shows that α-L-iduronate corresponds, in fact, to β-D-glucuronate).

Another important aspect of heparin structure concerns the position of the sulfate groups. On an average, the polysaccharide chains contain close to three sulfate groups per disaccharide; one of these is located on the amino group of glucosamine (although many heparin preparations are now known to contain high proportions - up to 30% - of *N*-acetylglucosamine (A3)), a second is bound to C-6 (or occasionally to C-3) of the glucosamine, and the third, usually present, is located exclusively on an iduronic acid residue. It appears that this interesting new feature of heparin structure may be intimately related to the biosynthesis of the iduronic acid units (see pp.295-297). The recent developments in our concepts of heparin structure are summarized in Fig. 53, which shows the structure of a tetrasaccharide containing a β-linked glucuronic acid residue and an α-L-iduronic acid unit sulfated in position 2.

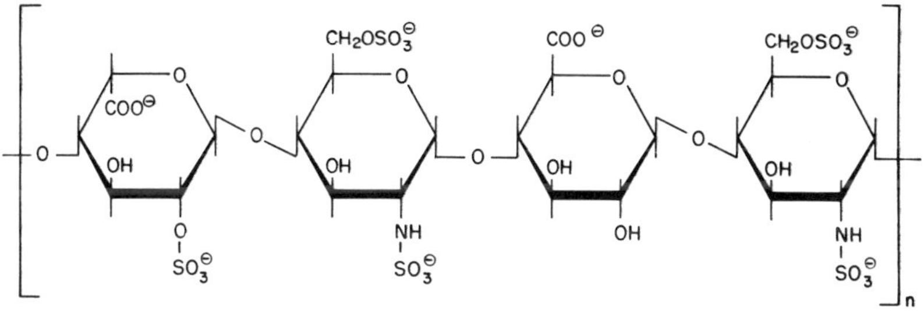

Figure 53 Structure of heparin tetrasaccharide containing glucuronic acid and 2-sulfated iduronic acid.

Heparan sulfate, like heparin, consists of alternating residues of uronic acid and glucosamine, that are sulfated to varying degrees. Heparan sulfate contains more N-acetyl groups and less sulfate than heparin; it generally lacks anticoagulant activity. As the difference in chemical structure between heparin and heparan sulfate is quantitative, and indistinct, rather than qualitative, the measurement of the anticoagulant activity is used to distinguish between these two mucopolysaccharides.

Dermatan sulfate, which until some years ago was thought to consist exclusively of repeating L-iduronic acid-containing disaccharide units, has now been shown, like heparin, to be a hybrid containing both iduronic acid and glucuronic acid, in which sulfate is sometimes present on the iduronic acid (although less frequently so than in heparin). As a result, the definition of dermatan sulfate must be somewhat revised; for example, this polysaccharide may no longer be regarded as entirely resistant to digestion with testicular hyaluronidase, since the presence of glucuronic acid renders the molecule partially susceptible to degradation by this enzyme.

STRUCTURE OF NATIVE PROTEOGLYCANS

The native cartilage proteoglycans are large macromolecules with many polysaccharide chains attached to the protein core. Thus, preparations obtained from bovine nasal septum by extraction with salt solutions have a molecular weight range of 1 to 4 million, corresponding to

40-150 glycosaminoglycan chains per molecule. Smaller proteoglycans occur in other tissues, such as pig laryngeal cartilage, in which the molecular weight ranges from about 60,000 to 1 million.

Little is known regarding the structure of the protein moiety of any of the proteoglycans. Considering that the protein content of proteoglycan from bovine nasal septum is approximately 7%, the molecular weight of the protein core is about 200,000. Amino acid analysis, proteolytic enzyme digestions and immunological studies of purified proteoglycan preparations yielded some information regarding the protein moiety of the proteoglycan, so called core protein. The core protein is composed of a single polypeptide chain of about 2,000 amino acid residues with alternating short and long repeating sequences. A short sequence consists of less than 10 amino acid residues with one N-terminal and C-terminal serine residue, each of which carries a polysaccharide chain linked glycosidically to its hydroxyl group. The long sequence contains about 35 amino acid residues. The main polypeptide chain is probably homologous in the vertebrate sub-phylum with strong conservation of structure suggested for the short sequence. However, polymorphism of polypeptide structures cannot be be excluded (A4).

The proteoglycans of cartilage and nucleus pulposus, which have been most extensively studied, contain both chondroitin sulfate and keratan sulfate as part of the same molecule. The relative proportion of each may vary, from molecules containing virtually no keratan sulfate, to those containing as much keratan sulfate as chondroitin sulfate. Recent evidence, mainly from the work of H. Muir in London(A5), V. Haskall in Ann Arbor, Michigan (A6) and L. Rosenberg in New York (A7), shows that proteoglycans of cartilage are capable of forming multiple aggregates, the aggregation depending upon the interaction of several proteoglycan molecules with a single hyaluronic acid molecule. According to a current model, based on chemical and electron microscopic studies, the proteoglycan aggregates are made up of an elongated filamentous hyaluronic acid backbone, 4,000 to 40,000 Å in length, from which proteoglycan molecules 1,000 to 4,000 Å long arise laterally at 200-300 Å intervals.

ENZYMIC DEGRADATION OF MUCOPOLYSACCHARIDES (7)

The structure of mucopolysaccharides, where a protein core holds together a number of carbohydrate chains, makes such a molecule particularly vulnerable to proteolytic breakdown; for this reason it would seem that during normal turnover the initial step would be an attack by proteolytic enzymes on the protein core. The most important proteases involved in the degradation of connective tissue proteoglycans are the cathepsins, a group of lysosomal enzymes of differing specificity.

The carbohydrate chains are probably first degraded to oligosaccharides by the action of endoglycosidases of rather narrow specificity and subsequently by specific exoglycosidases that remove sugar residues one at a time from the non-reducing end of the oligosaccharides formed by the endoglycosidases. So far, only one endoglycosidase has been identified in mammalian tissues, namely hyaluronidase. This is an endo-β-N-acetylhexosaminidase that degrades hyaluronic acid and both isomers of chondroitin sulfate, producing oligosaccharides having glucuronic acid at the non-reducing end. The pathway of degradation of dermatan sulfate, heparan sulfate and keratan sulfate, is still obscure.

REFERENCES

Reviews

1. Biochemistry and metabolism of the mucopolysaccharides,
    J. E. Silbert, Bull. Rheumatic Diseases $\underline{22}$, 680-685
    (1971-72).
    *An excellent introduction to the subject, though it
    may not be easy to locate the journal.*

2. Biochemistry and biology of mucopolysaccharides,
    K. Meyer, Amer. J. Med. $\underline{47}$, 664-672 (1969).
    A general survey by the pioneer of this field.

3. The structure and metabolism of mucopolysaccharides
    (glycosaminoglycans) and the problem of mucopoly-
    saccharidoses,
    H. Muir, Amer. J. Med. $\underline{47}$, 673-690 (1969).

4. Mucopolysaccharides of higher animals,
    R. W. Jeanloz *in* The Carbohydrates (Eds. W. Pigman,
    D. Horton and A. Herp) Vol.IIB, Academic Press, New
    York, 1970, pp. 589-625.
    More chemically oriented than references 1-3.

5. Carbohydrate-peptide linkages in proteoglycans of
    animal, plant and bacterial origin,
    U. Lindahl and L. Rodén *in* Glycoproteins (Ed. A.
    Gottschalk), 2nd ed., Elsevier, Amsterdam, 1972,
    pp.491-517.

6. Adventures in viscous solutions,
    A. Dorfman, Molecular and Cellular Biochem. $\underline{4}$, 45-65
    (1974).
    *The personal story of research on mucopolysaccharides
    by one of the central figures in the field. Covers
    both structural and metabolic aspects. Makes very
    good reading.*

7. Structure and enzymic degradation of mucopolysacchar-
    ides,
    H. Muir, *in* Lysosomes and Storage Diseases (Eds. H. G.
    Hers and F. van Hoof), Academic Press, 1973, pp.79-104.

Specific articles

A1. The linkage region of cartilage keratan sulfate to protein,
F. J. Kieras, J. Biol. Chem. 249, 7506-7513 (1974).

A2. The structure and composition of cartilage keratan sulphate,
J. J. Hopwood and H. C. Robinson, Biochem. J. 141, 517-526 (1974).
Proposes a structure for keratan sulfate-protein as found in skeletal proteoglycans.

A3. Structural studies on heparins with unusually high N-acetylglucosamine contents,
J. A. Cifonelli and J. King, Biochim. Biophys. Acta 320, 331-340 (1973).

A4. Comparative biochemistry of chondroitin sulphate proteins of cartilage and notochord,
M. B. Mathews, Biochem. J. 125, 37-46 (1971).

A5. Hyaluronic acid in cartilage and proteoglycan aggregation,
T. E. Hardingham and H. Muir, Biochem. J. 139, 565-581 (1974).

A6. Aggregation of cartilage proteoglycans. I. The role of hyaluronic acid,
V. C. Hascall and D. Heinegård, J. Biol. Chem. 249, 4232-4241 (1974).
See also the following paper (pp.4242-4249) by the same authors.

A7. Electron microscopic studies of proteoglycan aggregates from bovine articular cartilage,
L. Rosenberg, W. Hellmann and A. K. Kleinschmidt, J. Biol. Chem. 250, 1877-1883 (1975).

MUCOPOLYSACCHARIDES (PROTEOGLYCANS) -
II: BIOSYNTHESIS

The biosynthesis of proteoglycans (1,2) raises similar problems to those encountered in studies of the formation of other glycoproteins. One of the more obvious of these concerns the order in which the protein and polysaccharide moieties are synthesized. It has been found that compounds which inhibit protein synthesis specifically, such as puromycin and cycloheximide, likewise inhibit the incorporation of sugar and sulfate into proteoglycans. These observations are in accord with the idea that protein synthesis precedes the addition of the carbohydrate chains, as was found for the typical glycoproteins discussed earlier. It has not yet been established whether the glycosyl units are added to the completed protein or whether carbohydrate chain initiation occurs before the peptide is released from the ribosomes.

During biosynthesis of the polysaccharide chains, the proteoglycan moves through the smooth reticulum to the Golgi apparatus, as is the case for other glycoproteins. However, neither the exact location of the enzymes involved, nor the mode of excretion of the proteoglycan is known.

Interestingly it has recently been shown (A1) that xylose or β-xylosides, such as p-nitrophenyl-β-$\underline{D}$-xyloside, will overcome the puromycin inhibition of chondroitin sulfate synthesis in cultured chick embryo chondrocytes. These results indicate that under suitable conditions, synthesis of chondroitin sulfate chains may occur in the absence of the protein core.

In the following I shall discuss primarily the biosynthesis of chondroitin sulfate, of which we now have the clearest overall picture. Some of the more important problems concerning the other mucopolysaccharides will be discussed briefly.

BIOSYNTHESIS OF CHONDROITIN SULFATE

In accordance with the notion that growth of polysaccharide chains takes place by stepwise addition of single monosaccharide units from the respective sugar nucleotides, it can be postulated that the biosynthesis of chondroitin sulfate occurs by a chain-initiating xylosyl transfer to the protein core of the proteoglycan (or possibly to the growing peptide chain), followed by a series of additional glycosyl transfer reactions, eventually leading to the formation of a complete proteoglycan molecule.

At any given moment, a cartilage cell presumably contains polysaccharide chains in all stages of growth, and it is therefore not surprising that homogenates and subcellular fractions contain endogenous acceptors to which the various monosaccharide components of chondroitin sulfate can be transferred from their respective sugar nucleotides. Transfer to all four monosaccharide components - glucuronic acid, N-acetylgalactosamine, galactose and xylose - from their UDP-derivatives to endogenous acceptors, has indeed been demonstrated, and partial characterization of the products has shown that the incorporated sugars are part of the expected chondroitin sulfate structures.

However, much of our knowledge of the individual enzymes involved and of their specificity has emerged from the use of exogenous acceptors of well-defined nature.

XYLOSYL TRANSFER. This reaction was first demonstrated in 1966 in cell-free preparations of hen's oviduct, mouse mastocytoma, and chick embryonic cartilage, which catalyzed transfer of xylose from UDP-xylose to endogenous acceptors. It was shown that xylose had been incorporated into the protein by an alkali-labile linkage from which it could be released as xylitol on treatment with alkaline borohydride, and which, upon exhaustive proteolysis, yielded xylosyl-L-serine. Further information on the substrate specificity of the xylosyltransferase was obtained from studies with a variety of potential xylose acceptors ranging from serine and simple serine derivatives to the entire protein core of the chondroitin sulfate proteoglycan. It will be recalled that the serine-linked xylose residue has two vicinal hydroxyl groups (Fig.52) and it is therefore possible to cleave the xylosidic linkage between chondroitin sulfate and protein by the Smith degradation (see p. 91). Tryptic digests of chondroitin sulfate, as well as of the entire proteoglycan, were subjected to this procedure, and it was found that in both cases the products were xylose acceptors, the Smith-degraded proteoglycan being a far better acceptor than any of the smaller substrates. The observation that the macromolecular substrate had the highest acceptor activity is analogous to what has been observed in other instances of glycosyl transfer to polypeptide acceptors.

Interestingly, the intact cartilage proteoglycan did not serve as a xylosyl acceptor despite the fact that about half of its serine residues are not glycosylated. This could conceivably be the result of steric hindrance by the bulky chondroitin sulfate chains, but enzymic degradation products, in which the chondroitin sulfate chains had been reduced to short oligosaccharides by digestion with testicular hyaluronidase or bacterial chondroitinase, were also inactive. These observations lend support to the idea that the primary structure of the protein in the vicinity of the polysaccharide-substituted serine residues is of importance for xylosyl acceptor activity.

The presence of both chondroitin sulfate and keratan sulfate in the cartilage proteoglycan raises another intriguing problem concerning the substrate specificity of the glycosyltransferases. Some of the keratan sulfate chains are bound to serine, whereas others are linked to threonine, in both cases by glycosidic linkages to N-acetylgalactosamine; the chondroitin sulfate is linked solely to serine. Although it is conceivable that initiation of chondroitin sulfate and keratan sulfate chains could occur randomly at any serine site in the molecule, this does not seem likely in view of the findings mentioned above. It can rather be assumed that the protein core has a definite structure in which specific serine residues are substituted by chondroitin sulfate and others by keratan sulfate, while some are not utilized for glycosylation. It may thus be postulated that the xylosyltransferase specifically seeks out certain serine residues in the polypeptide, whereas an N-acetylgalactosaminyltransferase initiates keratan sulfate synthesis at other serine sites, as well as at threonine side chains.

GALACTOSYL TRANSFER. The chain initiating xylosyl transfer is followed by the sequential addition of two galactosyl units. The monosaccharide xylose is a good acceptor in a reaction representing the first galactose transfer, although xylosyl-serine is a better one.

$$\text{UDP-Gal} + \text{Xyl} \longrightarrow \text{Gal-}\beta\text{-}(1\rightarrow 4)\text{-Xyl} + \text{UDP}$$

For the second galactosyl transfer reaction, Gal-β-(1→4)-Xyl may be used as exogenous substrate:

$$\text{UDP-Gal} + \text{Gal-}\beta\text{-}(1\rightarrow 4)\text{-Xyl} \rightarrow \text{Gal-}\beta\text{-}(1\rightarrow 3)\text{-Gal-}\beta\text{-}(1\rightarrow 4)\text{-Xyl} + \text{UDP}$$

Glycosyltransferases are in general very specific and catalyze transfer only to a particular monosaccharide acceptor. However, this does not always hold true, and a notable exception is lactose synthetase which transfers galactose to either glucose or N-acetylglucosamine (p.23). It was therefore important to establish whether the two reactions above are catalyzed by one or two enzymes. As shown in mixed-substrate experiments, no competition occurred between acceptors for the two different transfer steps, e.g. xylose did not inhibit galactosyl transfer to Gal-β-(1→4)-Xyl and *vice versa*. It was therefore concluded that two different enzymes are involved in the incorporation of the two galactose residues to the linking region.

In contrast to the first galactosyltransferase, the second transferase absolutely requires a larger acceptor structure, with 4-*O*-substituted xylose as the penultimate sugar (e.g. Gal-β-(1→4)-Xyl). No transfer was observed to galactose or to Gal-β-(1→3)-Gal. Apparently, in this case the enzyme needs to recognize not only the terminal monosaccharide acceptor but also the penultimate sugar.

The higher degree of specificity in the second galactosyltransferase reaction clearly makes it impossible for the latter enzyme to add more than a single galactose residue to the galactosylxylose disaccharide unit. It is likely that this situation represents a common mechanism for the limited addition of identical units, although the substrate specificities of glycosyltransferases are not yet known in such detail as to permit definite generalizations in this regard.

GLUCURONYL TRANSFER TO GALACTOSE. The particulate fraction from a cartilage homogenate contains a glucuronyltransferase which catalyzes transfer to linkage region fragments such as Gal-β-(1$\rightarrow$3)-Gal:

UDP-GlcUA + Gal-β-(1$\rightarrow$3)-Gal $\longrightarrow$

GlcUA-β-(1$\rightarrow$3)-Gal-β-(1$\rightarrow$3)-Gal + UDP

Other galactose-containing compounds, including Gal-β-(1$\rightarrow$3)-Gal-β-(1$\rightarrow$4)-Xyl and Gal-β-(1$\rightarrow$3)-Gal-β-(1$\rightarrow$4)-Xyl-β-(1$\rightarrow$)-Ser, are also acceptors for glucuronyl transfer. The glucuronyltransferase reaction completes the formation of the specific linkage region of chondroitin sulfate.

REPEATING DISACCHARIDE FORMATION. Studies of the cell-free biosynthesis of chondroitin sulfate have shown that a low-sulfated polysaccharide was formed on incubation of a cell-free preparation from embryonic chick cartilage with UDP-glucuronic acid and UDP-N-acetylgalactosamine. Polymerization results from the concerted action of two glycosyltransferases, an N-acetylgalactosaminyltransferase and a glucuronyltransferase, which alternately add the two monosaccharide components of the repeating disaccharide units. As shown by A. Dorfman and his coworkers, N-acetylgalactosaminyl transfer to non-reducing terminal glucuronic acid takes place with sulfated as well as non-sulfated acceptor oligosaccharides from chondroitin, chondroitin 4-sulfate, chondroitin 6-sulfate, and hyaluronic acid. The finding that hyaluronic acid hexasaccharide is an acceptor shows that the transferase has a simple substrate specificity with a requirement only for the correct terminal monosaccharide acceptor, whereas the identity of the penultimate group is not important.

The glucuronyltransferase catalyzed glucuronyl transfer to oligosaccharides with a non-sulfated, or 6-sulfated N-acetylgalactosamine residue in the non-reducing terminal position. However, a 4-sulfated N-acetylgalactosamine unit does not serve as a glucuronyl acceptor. This suggests that, at least in the synthesis of chondroitin 4-sulfate, the addition of glucuronic acid precedes the sulfation of the N-acetylgalactosamine residue. Because of this specificity of the chain elongation enzymes, it has been suggested that sulfation at position 4 of the non-reducing N-acetylgalactosamine unit may serve as a mechanism of chain termination. No evidence is, however, available to support this notion. Use of these oligosaccharide acceptors does not lead to true polymerization, but only the addition of one or a few sugars. Use of endogenous material as acceptor or primer does, however, yield true polymerization.

A further brief comment should be made concerning the

relationship between the "polymerizing" glucuronyltransferase and the one catalyzing transfer to galactose in the linkage region. Competition experiments with Gal-β-(1→3)-Gal and a chondroitin 6-sulfate pentasaccharide with N-acetylgalactosamine at the non-reducing end indicated that two enzymes are indeed involved, since no competition occurred between these two substrates.

PARTICIPATION OF MULTIENZYME COMPLEXES

As we have seen, the assembly of a complete carbohydrate chain of a proteoglycan is a complex process. It has been estimated that some 10,000 separate steps are required for the synthesis of an intact proteoglycan molecule, with about 80 chondroitin sulfate side chains, each with a molecular weight of about 25,000. Obviously such a synthesis must occur in a highly organized way with the six glycosyltransferases working in concert, the product of one enzymic reaction becoming the substrate for and the determinant of the next reaction. This would require that the enzymes involved be closely located, arranged possibly in an orderly fashion on the membranes of the endoplasmic reticulum or Golgi region. Indeed, specific interaction between purified xylosyltransferase and the first galactosyltransferase that catalyse the initial reactions in the biosynthesis of chondroitin sulfate chains has now been demonstrated (A2). Whether such enzyme-enzyme interactions participate in the organization of a multienzyme glycosyltransferase complex *in vivo* remains to be established.

It should also be noted that the chondroitin sulfate glycosyltransferases are firmly associated with the membranes of the endoplasmic reticulum or the Golgi apparatus. This, incidentally, represents one of the major obstacles in the purification of these enzymes. Some progress has been made recently in this regard. I have just mentioned the xylosyltransferase and galactosyltransferase, which are two such enzymes that have been solubilized and purified.

As we have seen, six distinct glycosyltransferases seem to be involved in the synthesis of the polysaccharide chain of the chondroitin sulfates, and this process is summarized in Fig. 54. The sulfation of the polysaccharide, which is also indicated in the figure, is discussed below.

SULFATION. The completion of the chondroitin sulfate chains requires sulfation of the N-acetylgalactosamine residues in positions 4 or 6. Although biosynthesis of a sulfated sugar nucleotide (UDP-N-acetylgalactosamine-4-sulfate) has been described, it does not appear to be an intermediate in

MUCOPOLYSACCHARIDES II 293

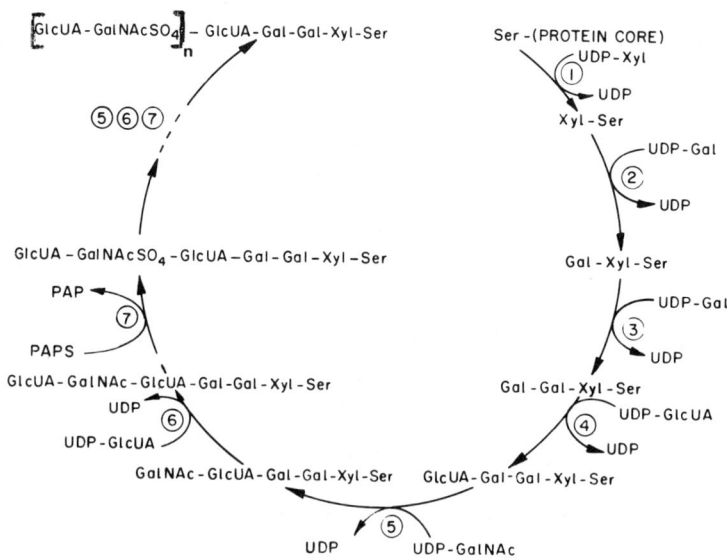

Figure 54 Enzymic steps in the biosynthesis of chondroitin 4-sulfate. The enzymes involved are xylosyltransferase [1], two galactosyltransferases [2],[3], glucuronyltransferase [4], N-acetylgalactosaminyltransferase [5], glucuronyltransferase [6] and sulfotransferase [7]. Enzymes 1-4 form the linkage region, while 5-7 catalyze the formation of the sulfated disaccharide repeating unit.

sulfation of mucopolysaccharides. Mainly due to the elegant work of J. E. Silbert at Boston (A3-A5), we now know that sulfation occurs on the polysaccharide, along with, or immediately after polymerization.

The sulfate donor in this reaction, as in other biological systems, is 3'-phosphoadenosyl-5'-phosphosulfate (PAPS), which is formed from ATP and sulfate in two steps, as shown originally by F. Lipmann and P. W. Robbins in Boston in 1956:

[1]   ATP + sulfate ⇌ AMP-sulfate + pyrophosphate
                      (adenylsulfate)

[2]   AMP-sulfate + ATP ⟶ 3'-phospho-AMP-sulfate + ADP
                                    (PAPS)

PAPS (like the sugar nucleotides) is synthesized by soluble enzymes in the cytoplasm. Direct transfer of a sulfate group from PAPS to both chondroitin sulfate and heparin has been demonstrated *in vitro*. In both cases, the particulate enzyme preparation that catalyses the polymerization from the sugar nucleotide precursors, also catalyses the incorporation of sulfate onto newly formed polysaccharide. Separation of sulfotransferase activity from the polymerases results in a much lower efficiency of sulfation of partially sulfated acceptors.

The biosynthesis of the N-sulfate groups in heparin and heparan sulfate has been shown by Silbert to occur, at least in part, in a rather unusual way, by displacement of N-acetyl groups on the preformed polymer. Using UDP-GlcNAc, labeled with tritium in the acetyl group, and UDP-GlcUA it was shown that in the absence of PAPS an N-acetylated polymer was formed by a particulate fraction from mastocytoma cells. In the presence of PAPS, however, there was a loss of as much as half the labeled acetyl groups and a corresponding formation of N-sulfate groups. This reaction took place equally well during or after polymerization. There is now evidence for the formation of a polymer with free amino groups as an intermediate between the N-acetylated and N-sulfated ones (A6). All the data suggest that the sulfating enzymes are located in a multienzyme complex, together with the polymerizing enzymes.

BIOSYNTHESIS OF L-IDURONIC ACID

In 1962 the epimerization of UDP-glucuronic acid to UDP-L-iduronic acid was described. The latter compound has naturally been assumed to be the precursor of polysaccharide-bound L-iduronic acid residues in heparin and dermatan sulfate, but such a role has never been demonstrated experimentally. In spite of this, interconversion of the monosaccharide components of glycoproteins and polysaccharides has been thought to occur exclusively at the sugar nucleotide level.

The finding that formation of L-iduronic acid occurs by epimerization not at the nucleotide level, but at the polymer level, was prompted by the discovery of such a modification in the biosynthesis of alginic acid (or alginate), a similar though unrelated polysaccharide (A7,3). Alginic acid is the major matrix polysaccharide of marine brown algae, and has also been found to be an extracellular product of certain bacteria, such as *Azotobacter vinelandii*. It is composed of two uronic acids, D-mannuronic and its 5-epimer, L-guluronic acid, which are structurally related in the same manner as D-glucuronic and L-iduronic acid (p.44).

D-Mannuronic acid *L-Guluronic acid*

During the 1960's it was generally believed that alginic acid is formed from the corresponding nucleoside diphospho-uronic acids. In 1971 A. Haug and B. Larsen, at the Norwegian Seaweed Research Institute in Trondheim, proved, however, that this belief was incorrect. From the culture media in which *Azotobacter vinelandii* was grown, they isolated an enzyme that was capable of converting an algal polymannuronic acid into a typical copolymer of mannuronic acid and L-guluronic acid. This product was indistinguishable from ordinary algal alginic acid. The conclusion drawn from these experiments was that a C-5 epimerization took place at the polymer level, resulting in the conversion of D-mannuronic acid units in the macromolecule to L-guluronic acid units. Interestingly, no reverse epimerization of L-guluronic acid units to D-mannuronic acid was observed.

The reaction described above was the first reported example of epimerization of sugar residues in a preformed polysaccharide molecule.

U. Lindahl and his coworkers at Uppsala, investigated the biosynthesis of iduronic acid in a cell-free system from mouse mastocytoma (A8). Using radioactive UDP-glucuronic acid as a precursor, incorporation of glucuronic acid into polymeric material could easily be demonstrated, but no trace of iduronic acid was detected in the product. As it had previously been shown that many of the iduronic acid residues of heparin are sulfated (whereas all the glucuronic acid is unsulfated), it was assumed that concomitant sulfation may be essential for the epimerization to iduronic acid. Indeed, when the system was supplemented with PAPS, 40-50% of the radioactivity incorporated from UDP-glucuronic acid into the polymer was present in iduronic acid.

Epimerization in the opposite direction, from L-iduronic acid to D-glucuronic acid, has now been detected in studies of the biosynthesis of dermatan sulfate. Stereochemical modification of glycosyl units in a polysaccharide chain does, therefore, appear to be of widespread occurrence.

LIPIDS IN MUCOPOLYSACCHARIDE BIOSYNTHESIS

The utilization of exogenous acceptors of well defined structure has been of great value in delineating the substrate specificity of each step in chondroitin sulfate synthesis. However, it must be remembered that the process *in vivo* differs from the artificial situation created in the study of the individual reactions. Although the results described above are consistent with a stepwise addition of each glycosyl unit to form the entire polysaccharide chain, there is still a possibility that unknown intermediates participate in this process. In particular, the occurrence of lipid intermediates of a polyprenol nature has not been definitely ruled out, even though a search for such intermediates has so far been unsuccessful. There is, however, some evidence that lipids which are prominent components of the membranes of the endoplasmic reticulum, are important for the normal activity of the glycosyltransferases involved in the biosynthesis of mucopolysaccharides.

REGULATION OF MUCOPOLYSACCHARIDE BIOSYNTHESIS

An important aspect that I have not previously discussed concerns the regulation of mucopolysaccharide metabolism. At the enzymic level two regulatory mechanisms have been described, by which the synthesis of the UDP-N-acetylhexosamines and UDP-glucuronic acid is controlled. The first specific reaction in hexosamine synthesis, i.e. the fructose 6-phosphate:glutamine transamidase reaction, is subject to feed-back regulation by UDP-GlcNAc, and since UDP-GlcNAc is in equilibrium with UDP-GalNAc, the formation of this precursor of chondroitin sulfate is also affected by the same mechanism (compare p.169, Fig.40).

Regulation of UDP-GalNAc synthesis is complemented by a second mechanism which affects the synthesis of UDP-glucuronic acid. E. F. Neufeld and C. W. Hall have shown that UDP-Xyl is a potent inhibitor of UDP-Glc-dehydrogenase, which oxidizes UDP-Glc to UDP-glucuronic acid. Since UDP-Xyl is formed by decarboxylation of UDP-glucuronic acid, the levels of both these nucleotide sugars will be regulated by a sensitive feed-back mechanism. It will be noted that UDP-Xyl may be regarded as a specific synchronizer of the synthesis of the protein and polysaccharide moieties of a proteoglycan, since the earliest effect of diminished protein synthesis on the carbohydrate precursors would be accumulation of UDP-Xyl. This would effectively decrease the synthesis of UDP-glucuronic acid and UDP-Xyl itself until the balance had been restored. The presence of xylose in the key position of the carbohydrate-protein linkage may thus be regarded as an expression of the need for a sensitive regulatory mechanism which would respond immediately to any imbalance during the biosynthesis of the protein and polysaccharide moieties of the conjugated macromolecule.

Another possible site of regulation of mucopolysaccharide biosynthesis at the enzymic level may result from the fact that out of the seven transferases involved in mucopolysaccharide biosynthesis (Fig.54), only the xylosyltransferase and the sulfotransferase are highly soluble, whereas the other enzymes are organized in a multi-particulate-complex. Disintegration and rejoining of the soluble enzymes might control the synthesis of the mucopolysaccharide molecules.

REFERENCES

Reviews

1. Biosynthesis of acidic glycosaminoglycans, L. Rodén *in* Metabolic Conjugation and Metabolic Hydrolysis (Ed. W. H. Fishman), Vol. 2, Academic Press, New York, 1970, pp.345-442.
    This is the most comprehensive review on biosynthesis, summarizing the subject as it was in 1970.

2. Some aspects of the structure and biosynthesis of connective tissue proteoglycans,
L. Rodén, J. R. Baker, N. B. Schwartz, A. C. Stoolmiller, S. Yamagata and T. Yamagata, *in* Biochemistry of the Glycosidic Linkage; An Integrated View (Eds. R. Piras and H. G. Pontis), PAABS Symposium Vol.2, Academic Press, 1972, pp.345-385.
An excellent review, more abbreviated and up to date than reference 1; much of the presentation in lectures 14 and 15 is based on this source.

3. Biosynthesis of algal polysaccharides,
A. Haug and B. Larsen, *in* Plant Carbohydrate Biochemistry (Ed. J. B. Pridham), Academic Press, 1974, pp.207-218.
Includes an interesting account of the epimerization of uronic acids at the polymer level, by the discoverers of this novel reaction.

Specific articles

A1. Stimulation of synthesis of free chondroitin sulfate chains by β-D-xylosides in cultured cells,
N. B. Schwartz, L. Galligani, P.L. Ho, and A. Dorfman, Proc. Natl. Acad. Sci. USA 71, 4047-4051 (1974).

A2. Biosynthesis of chondroitin sulfate. Purification of UDP-D-xylose: core protein β-D-xylosyltransferase by affinity chromatography,
N. B. Schwartz and L. Rodén, Carbohyd. Res. 37, 167-180 (1974).

A3. Biosynthesis of chondroitin sulfate. Sulfation of the polysaccharide chain,
S. DeLuca, M. E. Richmond and J. E. Silbert, Biochemistry 12, 3911-3915 (1973).

A4. Biosynthesis of chondroitin sulfate. Microsomal acceptors of sulfate, glucuronic acid, and N-acetylgalactosamine,
M. E. Richmond, S. DeLuca and J. E. Silbert, Biochemistry 12, 3898-3903 (1973).

A5. Biosynthesis of chondroitin sulfate. Assembly of chondroitin on microsomal primers,
M. E. Richmond, S. DeLuca and J. E. Silbert, Biochemistry 12, 3904-3910 (1973).

A6. Biosynthesis of heparin. II. Formation of sulfamino groups,
U. Lindahl, G. Bäckström, L. Jansson and A. Hallén,
J. Biol. Chem. 248, 7234-7241 (1973).

A7. Biosynthesis of alginate. Part II. Polymannuronic acid C-5-epimerase from *Azotobacter vinelandii* (Lipman),
A. Haug and B. Larsen, Carbohyd.Res. 17, 297-308 (1971).

A8. Biosynthesis of heparin. III. Formation of iduronic acid residues,
M. Höök, U. Lindahl, G. Bäckström, A. Malmström and L. Å. Fransson, J. Biol. Chem. 249, 3908-3915 (1974).

16

GENETIC DEFECTS OF MUCOPOLYSACCHARIDE METABOLISM

Disorders in the metabolism of mucopolysaccharides in man have been recognized for some time, and it was found quite early that these diseases, known collectively as "mucopolysaccharidoses", are of genetic origin. However, the understanding of the biochemical basis of these disorders had to await the characterization of the different mucopolysaccharides. No wonder, therefore, that the enzymic defects of most mucopolysaccharidoses have been elucidated only within the past few years (1-4). This was mainly due to the efforts of E. F. Neufeld and her collaborators at the National Institutes of Health, and A. Dorfman and his group at the University of Chicago.

The best known of the mucopolysaccharidoses are the Hurler and Hunter syndromes, first described in detail some

60 years ago. These are recessively inherited diseases, characterized by skeletal abnormalities and mental retardation, which in severe cases result in early death. A major difference between the Hurler and Hunter syndromes is in the mode of inheritance. The Hurler syndrome is transmitted in classical Mendelian fashion as an autosomal recessive; it can occur in children of either sex whose parents, though carriers of the Hurler gene, show no apparent abnormality. The Hunter syndrome is sex linked, like hemophilia for example. Women who are carriers can transmit the disease to their sons but not to their daughters; half of the daughters, however, are likely to be carriers and in turn transmit the disease to their sons. Another disorder, the Sanfilippo syndrome, resembles both the Hurler and Hunter syndromes, except that the physical defects are relatively mild, while mental retardation is severe. It is transmitted, like the Hurler syndrome, as an autosomal recessive.

The frequency with which these and other closely related syndromes occur is not precisely known. There are some statistics that give 1/100,000 live births for the Hunter syndrome in the U.S.; it appears that the collective incidence for all mucopolysaccharidoses is one per 30,000 births. In the United States there are at least several hundred families who have afflicted children. The diseases are not peculiar to any ethnic or racial group, and case reports from Europe suggest an incidence not unlike that in the United States.

Diagnosis of these disorders usually depends on the finding of elevated mucopolysaccharide excretion in the urine. About 100 mg of mucopolysaccharide, consisting primarily of low molecular weight dermatan sulfate and heparan sulfate, are excreted by Hurler and Hunter patients, compared to a normal excretion of 10 mg per day or less for all mucopolysaccharides combined. Sanfilippo patients excrete mainly heparan sulfate. In addition to urine, the mucopolysaccharides accumulate in many tissues. The liver, for instance, tends to contain mainly heparan sulfate, while dermatan sulfate is predominant in the spleen. It is the damage wrought by the cellular deposits of the mucopolysaccharides that probably underlies most, if not all, of the clinical problems.

Two related diseases are the Maroteaux-Lamy syndrome which involves dermatan sulfate, and the Morquio syndrome. The latter is unusual in that mental development is normal, whereas the skeletal deformities are associated with the excretion of keratan sulfate, a mucopolysaccharide different from dermatan sulfate and heparan sulfate (p.272).

In the connective tissues of patients with mucopolysaccharide disorders, one finds, as in normal individuals, the large proteoglycan entities of dermatan sulfate and heparan sulfate. However, in the urine or in tissues such as liver, where abnormal storage takes place, the molecules of these two mucopolysaccharides are much smaller. The protein core is missing, and only a few amino acids are attached to some of the carbohydrate chains, while other fragments have neither amino acids nor the xylosyl-serine linkage region; it is as if the protein portion had been fully digested while the polysaccharide had been only partially cleaved.

The first suggestion that a disorder of degradation might be the basis of these diseases was made by F. van Hoof and G. H. Hers in 1964 from the examination of the ultrastructure of liver from Hurler patients. The liver cells were full of vacuoles delineated by single membranes, containing very finely dispersed material. The vacuoles were thought to be lysosomes, the cellular organelles in which breakdown of macromolecules normally takes place, pathologically engorged with undigested mucopolysaccharides.

Lysosomes filled with mucopolysaccharides have also been found in the livers of Hunter and Sanfilippo patients. The mucopolysaccharidoses are now recognized as belonging to the large group of inherited disorders of lysosomal metabolism or the lysosomal diseases.

The structure of the stored and excreted polysaccharides suggested that in these disorders the mucopolysaccharides are not fully degraded to small units which could be returned to the general metabolic pool. This immediately raised some obvious questions: are the mucopolysaccharides chemically faulty? Does the affected individual produce them in such quantities as to overwhelm the normal disposal machinery? Is there a defect in the degradative mechanism? Answers to these questions had to come from biochemical studies of isolated tissues. In the absence of animals with similar genetic disorders, such studies became feasible only in recent years with the introduction of tissue and cell culture techniques.

CELL CULTURE STUDIES

A turning point in the study of mucopolysaccharidoses, as of so many other diseases, occurred in 1965 when it was reported that fibroblasts cultured from skin of patients with the Hurler or Hunter syndromes accumulate the two mucopolysaccharides, dermatan sulfate and heparan sulfate. Since skin biopsies are readily available and fibroblasts are easy to grow, lines of genetically marked cells could be used for metabolic study of the mucopolysaccharide disorders. Moreover, because the cells can be shipped around the world or stored frozen, research on the mucopolysaccharidoses became independent in time and place from the medical care of the patients.

Using cell cultures, E. F. Neufeld and her coworkers investigated whether accumulation of mucopolysaccharides in fibroblasts was due to increased rate of synthesis or to a decreased rate of degradation. They found that normal cells exposed to medium containing radioactive sulfate synthesize radioactive mucopolysaccharide, the fate of which can readily be followed. Once synthesized, about three quarters of the mucopolysaccharide is secreted by the fibroblasts into the medium, while the remainder is diverted to a storage pool within the cell, from which the only exit is by degradation. In Hurler and Hunter fibroblasts, material is admitted into that pool at normal rates, but is degraded relatively slowly. Thus in normal cells, most of the mucopolysaccharides have a half-life of about 6 hours, while a minor fraction has a half-life of about 3 days. In Hurler and Hunter cells, there is little degradation in the first 8 hours; the half-life of the entire storage pool varies, in different cultures, from 2 to 6 days. It appeared, therefore, that there may be two pools of degradation in normal cells, the faster of which is missing in the cells affected by the mucopolysaccharide disorders.

Superficially, no significant differences in the pattern of mucopolysaccharide metabolism in Hurler and Hunter fibroblasts are observed, just as it is impossible to distinguish between the two syndromes by examining the urinary mucopolysaccharides. If, however, fibroblasts from Hurler and Hunter patients are cultured together, the defects are corrected and the mixed cells show a normal pattern of metabolism. The presence of both cell types in the same dish allows each one to metabolize mucopolysaccharide normally. The basis for this correction turns out to be very simple. Because the genes responsible for the two disorders are different, the fundamental biochemical defects - that is, the proteins whose structure is encoded in those genes - must differ, however similar the end results might be. Hurler cells are normal with respect to the Hunter defect, and *vice versa*. The fibroblasts of one genotype are able to supply, through the medium, a factor lacking in cells of the second genotype.

As expected, the ability to correct the defect of Hurler cells is not limited to Hunter cells but is a property of other fibroblasts, provided they are not of the Hurler genotype. There are some exceptions, e.g. fibroblasts from patients with the Scheie syndrome, which is biochemically a variant of the Hurler syndrome. The metabolism of Hunter cells is similarly corrected by fibroblasts of all genotypes tested except those of the Hunter genotype. The defect in Sanfilippo fibroblasts can be similarly corrected.

IDENTIFICATION OF THE MISSING ENZYMES

Correction of the aberrant pattern of mucopolysaccharide metabolism is achieved not only by intact fibroblast, but also by the medium in which the cells were grown, by homogenates of fibroblasts or connective tissues, and even by factors extracted from normal urine.

The active substances or "corrective factors" affecting normalization of mucopolysaccharide catabolism were only very recently extensively purified from normal urine, and eventually shown to be the missing enzyme in each disorder (Table 16). Thus, the corrective factor for Hurler and

Table 16

Mucopolysaccharidoses

Name	Accumulated product	Enzyme deficiency
Hurler, Scheie	Heparan sulfate Dermatan sulfate	α-L-Iduronidase [2]
Hunter	Heparan sulfate Dermatan sulfate	Iduronate sulfatase [1]
Sanfilippo A	Heparan sulfate	Heparan sulfamidase [5]
Sanfilippo B	Heparan sulfate	α-N-Acetylglucosamin- idase [6]
Morquio	Keratan sulfate Chondroitin sulfate	N-Acetylhexosamine 6-sulfatase
Maroteaux-Lamy	Dermatan sulfate	N-Acetylgalactosamine 4-sulfatase [3]
Mucopolysacchar- idosis VII	?	β-Glucuronidase [4]

Dermatan sulfate: $\underline{L}\text{-IdUA} \xrightarrow{[2]}_{\alpha} \text{GalNAc} \xrightarrow{}_{\beta} \text{GlcUA} \xrightarrow{[4]}_{\beta} \text{GalNAc} \xrightarrow{}_{\beta}$
with [1] OSO_3^- on L-IdUA, [3] OSO_3^- on GalNAc, and OSO_3^- on second GalNAc

Heparan sulfate: $\underline{L}\text{-IdUA} \xrightarrow{[2]}_{\alpha} \text{GlcN} \xrightarrow{}_{\alpha} \text{GlcUA} \xrightarrow{[4]}_{\beta} \text{GlcNAc} \xrightarrow{[6]}_{\alpha}$
with [1] OSO_3^- on L-IdUA, [5] SO_3^- on GlcN, and OSO_3^- on GlcNAc

Summary of the enzymic defects in the mucopoly-saccharidoses. The numbers in brackets refer to the enzymes listed above.

Scheie cells is the enzyme α-L-iduronidase (enzyme [2] in Table 16), the Hunter corrective factor - iduronate sulfatase [1], the Sanfilippo A and B corrective factors - heparan N-sulfatase (heparan sulfamidase)[5] and α-N-acetylglucosaminidase [6], respectively. β-Glucuronidase [4], of human or bovine origin, serves as corrective factor for cells deficient in that enzyme. Identity of the Maroteaux-Lamy factor with arylsulfatase B or N-acetylgalactosamine 4-sulfatase [3], and of the Morquio factor with the corresponding 6-sulfatase, is probable but definitive proof must await the preparation of these enzymes in pure form.

Correction is accompanied by uptake of enzyme from the medium into fibroblasts, presumably into lysosomes. Normal catabolic function is thus restored, since lysosomes contain not only stored mucopolysaccharides, but also all other enzymes required for the degradation of these polymers. Only a small fraction of the normal complement of enzyme needs to be incorporated in order to give essentially normal mucopolysaccharide catabolism.

The uptake of lysosomal enzymes, originally thought to take place by pinocytic imbibition of medium with all its macromolecular contents, turned out to be a highly selective process requiring a recognition marker on the protein and, by implication, a receptor on the fibroblast surface. Preliminary evidence suggests that the recognition marker is a carbohydrate, since it is readily destroyed by periodate.

In the conceptual framework of degradation by exoenzymes, a block in the degradative pathway should lead to the accumulation of macromolecules with a non-reducing terminal residue that would have been the substrate for the missing enzyme. This prediction has been experimentally verified for the mucopolysaccharides that accumulate in fibroblasts derived from patients with the Hunter or the Maroteaux-Lamy syndrome, or with β-glucuronidase deficiency. The stored polymers have terminal sulfated iduronic acid, sulfated N-acetylgalactosamine and β-linked glucuronic acid residues, respectively.

Although disorders of chondroitin sulfate metabolism have been reported, they seem to be much rarer than the disorders of dermatan sulfate and heparan sulfate, which we have discussed at length. Also, it is remarkable that despite the close structural relationship between heparin and heparan sulfate there is no evidence that heparin metabolism is abnormal in the mucopolysaccharidoses.

Another point of interest is that no abnormalities in

the synthesis of proteoglycans have yet been found in any of the mucopolysaccharidoses. This is probably because so few genetic diseases have yet been investigated in depth. There is a number of genetic malformations which are likely candidates for disorders in synthesizing enzymes ("synthetic disorders").

PRACTICAL BENEFITS

The pattern of mucopolysaccharide metabolism by cells affected by the different mucopolysaccharidoses is so strikingly different from the normal that it can be used for prenatal diagnosis, a situation in which clinical observation is obviously impossible. The fetus is constantly shedding cells into the amniotic fluid, a sample of which can be withdrawn as early as 14 weeks after conception. Of the many cell types originally present in the fluid, fetal fibroblasts are the only ones to multiply in culture. If the fetus is affected with any of the mucopolysaccharidoses, such fibroblasts show the same abnormalities as fibroblasts from skin biopsies of affected adults. It is therefore possible to diagnose the disease either by examination of mucopolysaccharide accumulation or by enzyme tests. Prenatal diagnosis of the mucopolysaccharidoses has been of utmost value for families who have children affected with one of these diseases, since in the case of a subsequent pregnancy it provides a way for testing whether the fetus is normal or not.

The discovery of the biochemical defects in the various lysosome deficiency diseases was accompanied by the expectation and hope that it would be possible to treat the affected patients by replacing the missing enzymes. Attempts of enzyme replacement therapy made during the past ten years have clarified many problems that must be overcome and have suggested some general approaches.

One possible approach is to supply the patient with blood plasma cells or tissues that would serve as a continuous source of the enzyme. Infusion of leucocytes into patients with mucopolysaccharidoses has been tried, but the effects were transient and variable.

The present emphasis is on testing highly purified enzymes. None of the enzymes which are deficient in the mucopolysaccharidoses have yet been available in the large quantities required for injection into patients. A limited number of experiments along this line has been carried out on patients with lipidoses. For example, following injection of glucosylceramide β-glucosidase to a patient with the adult

form of Gaucher's disease, a relatively long-lasting reduction of the accumulated β-glucosylceramide was observed.
In anticipation that such an approach will eventually prove beneficial, at least in selected cases, procedures are being developed for the stabilization of lysosomal enzymes by chemical alteration, and for the decrease or abolition of their immunogenicity by encapsulation in non-antigenic material such as the recipients' own erythrocyte membranes.

Another promising approach, which I mentioned earlier (p.203), involves the attachment to enzymes of saccharide side chains. These may serve as recognition markers and direct the enzymes to the affected cells, in a manner similar to the function of exposed galactose residues on ceruloplasmin and other serum glycoproteins (see lecture 11).

REFERENCES

1. Inborn errors of mucopolysaccharide metabolism, E. F. Neufeld and J. C. Fratantoni, Science 169, 141-146 (1970).
   A basic review article on the subject, although somewhat outdated. Includes a brief and lucid discussion of the biochemistry of mucopolysaccharides.

2. The mucopolysaccharidoses, A. Dorfman and R. Matalon, *in* The Metabolic Basis of Inherited Disease (Eds. J. B. Stanbury, J. B. Wyngaarden, and D. S. Fredrickson), 3rd edit., McGraw Hill Book Co., 1972, pp.1218-1272.

GENETIC DEFECTS   317

3. The biochemical basis for mucopolysaccharidoses and
    mucolipidoses,
    E. F. Neufeld, *in* Progress in Medical Genetics (Eds.
    A. G. Steinberg and A. G. Bearn), Vol.10, Grune and
    Stratton, New York, 1974, pp.81-101.
    An up to date survey.

4. Adventures in viscous solutions,
    A. Dorfman, Molecular and Cellular Biochem. $\underline{4}$, 45-65
    (1974).
    *Includes descriptions of recent work, mainly from
    the author's laboratory, on mucopolysaccharidoses.*

17

THE BACTERIAL CELL WALL

A course on complex carbohydrates cannot be considered complete if it does not include a discussion of at least some of the special polymers of this class found in sources other than higher organisms. Most prominent among these are the diverse and highly complex polysaccharides located in the bacterial cell wall or envelope that surrounds the fragile cytoplasmic membrane of the bacterial cell (1-4). The cell wall, which is rigid, protects the membrane and the cytoplasm within from the adverse effects of the environment, and is thus responsible for the resistance of the bacterial cell to mechanical and osmotic injury. Moreover, it is the structure that provides the bacterial cell with its characteristic shape, whether spherical (a coccus) or rod-like (a bacillus). The cell wall is thus one of the features distinguishing bacteria from animal cells, which are enclosed only by a plasma membrane.

Most bacterial cells are 0.5 - 1 µ wide and 1 - 5 µ long; the thickness of the wall may vary between 80 and 200 Å, and its weight may account for up to 20 or 30% of the dry weight of the bacterial cell. Since the cell wall is insoluble in water (or in fact in any other solvent) it can be easily isolated after mechanical disintegration of the bacteria, followed by differential centrifugation to remove the soluble cytoplasmic constituents and the membrane fragments which do not sediment. The final product, when examined under the electron microscope, looks like a collapsed sac or balloon, with the same shape as the bacteria from which the walls were derived: the isolated cell walls of cocci are round and the walls of bacilli are elongated.

The foundations for the investigation of the bacterial cell wall were laid down by M. R. J. Salton at the University of Manchester in the early 1950's. One of the first generalizations to emerge from the early work on this subject was that with respect to their composition, structure and morphology, cell walls can be divided into two classes, similar to the old time division of bacteria according to the Gram stain (Table 17 and Fig.55).

Table 17

Major constituents of bacterial cell walls

	Gram-positive	Gram-negative
Peptidoglycan	+	+
Teichoic acids	+	−
Lipopolysaccharides	−	+
Other polysaccharides	+	±
Proteins	±	+
Lipoproteins	−	+

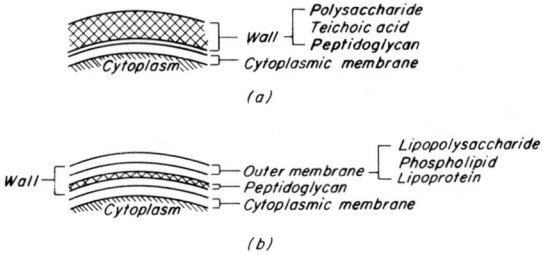

Figure 55 Schematic representation of the bacterial cell wall. (a) Wall of Gram-positive bacteria; (b) envelope of Gram-negative bacteria.

As can be seen, there are marked differences in the composition of the cell walls of Gram-positive and Gram-negative bacteria. Walls of both classes of bacteria do contain, however, one common constituent, peptidoglycan, which I shall discuss in detail in the last two lectures of the course. Suffice it to say at this point that peptidoglycan, also known as glycopeptide, mucopeptide or murein, is a unique type of polymer. It consists of linear polysaccharide strands crosslinked by oligopeptide units to form a single giant bag-shaped macromolecule, the molecular weight of which may be as high as 5×10^{10}. Peptidoglycan is the component responsible for the most basic function of the cell wall, its rigidity. It is resistant to the action of the common proteases or polysaccharidases, but is hydrolysed by lytic enzymes such as lysozyme or lysostaphin. Digestion of peptidoglycan by these enzymes, or interference in its synthesis by antibiotics such as penicillin, leads to lysis of the bacterial cell.

The various non-peptidoglycan components of bacterial cell walls may appear as useless accessories. Actually, owing to their external location on the surface of the cells, they are the sites of important biological activities. For example, they are antigens, or at least contain the chemical determinants responsible for the antigenic specificity of the bacteria. They contain the specific receptors which mediate phage fixation. Some are powerful toxins, while others seem to be implicated in the virulence of bacteria. Autolytic enzymes associated with the wall are believed to participate in growth of the wall and in separation of daughter cells.

Considerable variations in the composition of cell walls occur when bacteria are grown under different conditions. It now seems that the wall may be one of the most dynamic and phenotypically variable structures of the cell. Nevertheless, the basic organization of the peptidoglycan and most of its essential structural features are constant.

GRAM-POSITIVE BACTERIA

In walls of Gram-positive bacteria, peptidoglycan is the major constituent. In certain organisms, such as *Micrococcus luteus* (the organism first isolated in 1922 by A. Fleming, and known until recently as *Micrococcus lysodeikticus*), peptidoglycan accounts for about three-quarters of the dry weight of the wall. Walls of Gram-positive bacteria also contain one or more heteropolysaccharides, which are usually attached by covalent linkages to the glycan strands of the peptidoglycan. Most prominent among these are the teichoic acids, discovered in the 1950's by J. Baddiley and his coworkers in Newcastle-upon-Tyne, and thoroughly investigated by his group since then (5,6). The teichoic acids are polymers of ribitol phosphate or glycerol phosphate linked by phosphodiester bonds. In these polymers, the free hydroxyl groups of the ribitol or glycerol are often substituted by glycosidically linked monosaccharides (such as glucose or N-acetylglucosamine) or oligosaccharides (e.g. kojibiose, Glc-α-(1$\rightarrow$2)-Glc), as well as by an amino acid, $\underline{D}$-alanine, bound by an ester linkage. Polymers containing glycerol or ribitol that do not conform to the above structural pattern are also present in Gram-positive bacteria. Many of these teichoic acid-like polymers contain reducing sugars as part of the main chain. In polymers of this type, the repeating units may be joined together by phosphodiester linkages between the C-1 of glycerol and C-4 or C-6 of a hexose on the one hand, and the C-1 of the hexose and C-3 of glycerol on the other (Fig.56).

In Gram-positive bacteria, teichoic acids frequently account for between 30 and 50% of the dry weight of the wall. Although originally believed to be confined to cell walls, glycerol teichoic acids are now known to occur in bacterial membranes as well. The role of teichoic acids in the wall or membrane is still unknown. Many teichoic acids are serologically active and often constitute the group-specific components of Gram-positive bacteria. They also serve as receptors for phages. It has been suggested that they are

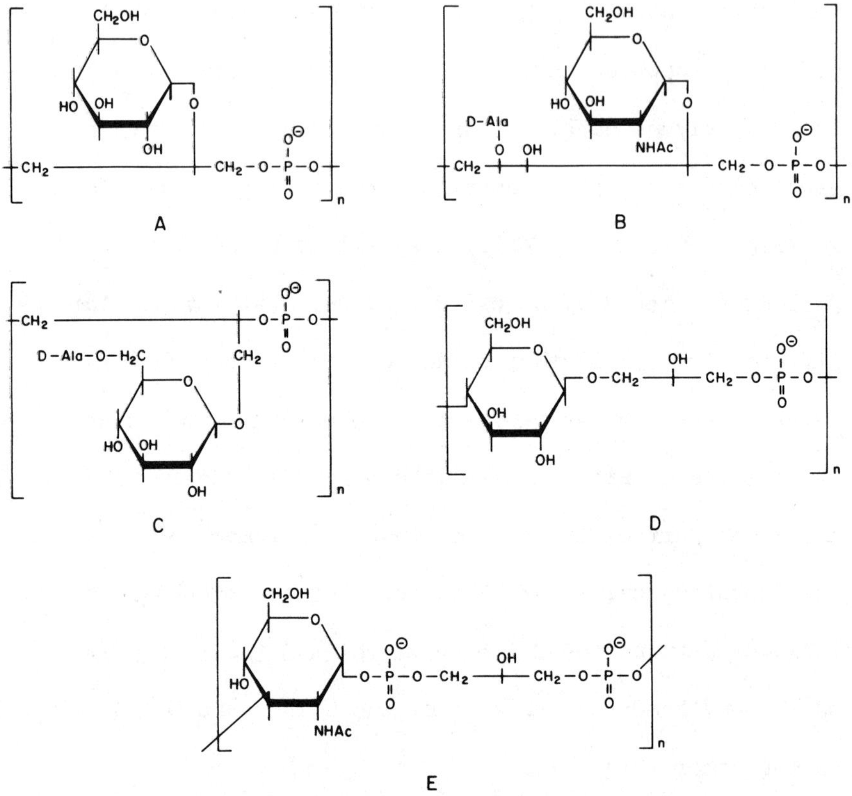

Figure 56 Some representative teichoic acid structures (modified from reference 3).

A. Poly-1,3-glycerol phosphate. The secondary hydroxyl group may be free, or glycosylated, or esterified by D-alanine. This is the most common type of teichoic acid structure. It is found in all membrane teichoic acids and in the cell wall teichoic acids of many Gram-positive organisms, for example *Staphylococcus*, *Streptococcus*, *Lactobacillus*, and *Bacillus* species.

B. Poly-1,5-ribitol phosphate. This structure is also common in the cell wall teichoic acids of the genera mentioned above. Glycosylation occurs on the 4-hydroxyl of ribitol, and the 2-hydroxyl is esterified by D-alanine, as shown.

C. Poly-1,2-glycerol phosphate. This polymer, substituted on its primary hydroxyl groups by 6-*O*-D-alanyl-glucose as indicated, has been found in *Bacillus stearothermophilus*.

D. Poly-4-phospho-β-glucosyl-glycerol. This polymer and its galactose analog have been found in *Bacillus licheniformis* and in *Lactobacillus plantarum*. Polymers with this type of structure have been found in pneumococcal capsular polysaccharides and pneumococcal C-substance.

E. This is a member of the less common type of teichoic acid in which the main chain contains the acid-labile sugar 1-phosphate linkage. This example is from *Staphylococcus lactis* I3.

involved in binding of cations (in particular Mg^{++}) to the cell surface, but this has not yet been established with certainty.

It has been known for a long time that all strains of *Diplococcus pneumonia*, the causative agent of pneumococcal pneumonia, possess a common polysaccharide antigen, the species specific C-substance. D. E. Brundish and J. Baddiley have shown (A1) that this polysaccharide contains ribitol phosphate and therefore should be considered as a teichoic acid. In addition to ribitol phosphate, the pneumococcal C-polysaccharide contains galactosamine-6-phosphate (which is probably the determinant of serological activity), glucose, choline phosphate (in its repeating structure), as well as a 2,4-diamino-2,4,6-trideoxy hexose, the structure of which was not elucidated. The latter may be identical with the diamino sugar which I isolated some 15 years ago from a polysaccharide of *B. licheniformis*, and which we recently identified through degradation and synthesis as 2,4-diamino-2,4,6-trideoxy-D̲-glucose (see p.18).

Pneumococci also produce type specific polysaccharides that play an important role in the virulence of this organism (7). These polysaccharides are external to the cell wall, and form the capsules that cover the pneumococcal cells. They are soluble in water, and can therefore be removed from the wall by washing. Some 75 different polysaccharides have been identified, and the structures of many of these have been worked out. Some of the pneumococcal polysaccharides, such as types 13 and 34, contain ribitol phosphate in addition to other sugars (e.g. glucose, galactose and N-acetylglucosamine in type 13, glucose and galactose in type 34) and are therefore considered as teichoic acids.

Studies on the pneumococcal polysaccharides, besides their value in the development of methods for immunization against pneumonia, had an enormous impact on immunology, genetics and modern biology. These polysaccharides were the first non-protein materials shown to be antigenic, as demonstrated by Avery, Heidelberger and Goebel at the Rockefeller Institute in the 1920's (see pp.26-7). Moreover, the revolutionary work on bacterial transformation, in which Avery,

MacLeod and McCarthy in 1944 identified DNA as the genetic material, was done with pneumococci. And as you may recall, the genetic marker was the production of a specific capsular polysaccharide! Unfortunately, because of shortage of time, I shall not be able to elaborate on this fascinating topic.

Cell wall polysaccharides are the group specific antigens of the *Lactobacilli* and the pathogenic *Streptococci*. For example, the group specific antigen in *Streptococci* of group A is a polysaccharide which consists of $\underline{L}$-rhamnose and N-acetylglucosamine; in group B the specific polysaccharide is made up of glucose, galactose and $\underline{L}$-rhamnose, and in group C of $\underline{L}$-rhamnose and N-acetylgalactosamine. However, in D and N strains, membrane glycerol teichoic acid is the group-specific antigen. The strain-specific antigens in group A strains are the M proteins which, together with the hyaluronate capsule, determine the virulence of these organisms. The M proteins are probably covalently linked to peptidoglycan. Other proteins, called T and R proteins, are also present in these walls. While these proteins do not contribute to the virulence of the *Streptococci*, they may contain the antigenic determinants which appear to cause the development of autoimmune diseases such as rheumatic heart disease or chronic glomerulonephritis, that result from streptococcal infections.

Other polysaccharides have been isolated from walls of Gram-positive bacteria. An interesting example are the teichuronic acids, polymers made up from alternating units of a uronic acid and a hexose or hexosamine. In walls of *M. luteus* the teichuronic acid (the only component besides the peptidoglycan) consists of glucose and N-acetylaminomannuronic acid, while the teichuronic acid of *B. licheniformis* is made up of glucuronic acid and N-acetylgalactosamine (note the similarity of the latter polymer to chondroitin and to hyaluronic acid).

GRAM-NEGATIVE BACTERIA

Cell walls of Gram-negative bacteria are considerably more complex than those of Gram-positive cells. The structure surrounding Gram-negative bacteria is more properly described by the term "cell envelope". In these organisms a thin peptidoglycan layer (most likely a single monomolecular layer), 20-30 $\overset{\circ}{A}$ thick, is sandwiched between the cytoplasmic membrane and an outer membrane-like structure. Such a triple layer can be seen upon examination in the electron microscope of sections of intact Gram-negative bacteria. The outer layer contains protein, phospholipid, lipopolysaccharide and lipoprotein. As a result, the peptidoglycan is not accessible to the action of lysozyme or of other lytic enzymes unless this layer is damaged, either physically (by freezing and thawing the cells) or chemically (by exposure to dilute alkali or to a metal-chelating agent such as EDTA). The outer membrane of Gram-negative bacteria also acts as a permeability barrier to many antibiotics, such as various penicillins, vancomycin and bacitracin, as well as the macrolide antibiotics (e.g. erythromycin), and is thus largely responsible for the resistance of these bacteria to the above agents.

The lipopolysaccharide, or LPS, does not appear to be covalently linked to the peptidoglycan, or trapped within this layer, since the bulk of it can be extracted by reagents, such as phenol or EDTA, that would not be expected to break covalent bonds. The conclusion that it is located on the outer surface of the bacterial cell is based among other things on the finding that the immunological specificity of the intact bacteria is, as a rule, the same as that of the purified LPS. Indeed, the purified LPS represents the so-called O-antigen of Gram-negative bacteria, such as the *Salmonella, Shigella* and *E. coli*. Moreover, purified LPS will bind bacteriophages such as the T even phages of

E. *coli* or the ε phages of *Salmonella*, with the same specificity as the intact cells from which the LPS was obtained.

In isolated cell walls of Gram-negative bacteria, LPS comprises 20-30% of the dry weight, while the peptidoglycan content is only 5-10% of their weight.

Walls of Gram-negative bacteria prepared under mild conditions contain the three layers of inner membrane, peptidoglycan and outer membrane. By extracting such walls with phenol-water (1:1) at 68°C (a technique also used for isolation of nucleic acids, but originally developed by O. Westphal for the isolation of LPS), the LPS and protein dissolve whereas the peptidoglycan remains insoluble, without change in its shape. Upon cooling the phenol-water mixture, the two phases separate, with the protein remaining in the phenol phase and the LPS in the aqueous phase. The LPS can be collected by high speed centrifugation, since it forms aggregates of very high molecular weight, up to several millions. LPS can also be prepared by phenol-water extraction of intact bacteria.

Another method of preparation of LPS is by extraction of cell walls with sodium dodecyl sulfate (SDS). U. Schwarz and V. Braun in Germany have found that treatment of cell walls of *E. coli* with SDS dissolves the outer and inner membranes, leaving the peptidoglycan ("murein sacculus") to which a lipoprotein is covalently bound (8). This lipoprotein is the major protein of the outer membrane of *E. coli* and related enterobacteriaceae, and also occurs free in the cell envelope, in an amount exceeding the bound form by a factor of two. It is an unusual macromolecule, made up of a single polypeptide chain of 58 amino acids which contains at its N-terminal end a covalently linked lipid. The C-terminal lysine of the lipoprotein is linked by a peptide linkage to a free amino group (of diaminopimelic acid) in the cell wall peptidoglycan. In a major part of the polypeptide chain, every third or fourth amino acid residue is hydrophobic. Electron micrographs indicate that the lipoprotein is attached to the outer surface of the peptidoglycan, where it probably interacts with, and binds, the components of the outer membrane through hydrophobic and other non-covalent bonds.

THE BIOLOGICALLY VERSATILE LIPOPOLYSACCHARIDES (9,10)

LPS have been the object of intensive biological investigations for many years, for several important reasons. Two of these I have mentioned earlier: LPS are immunologically active and they often serve as receptors for bacteriophages. Another interesting property of LPS is their pharmacological activity and toxicity for higher animals. Most remarkable is the pyrogenicity of LPS: injection of 1 µg of purified LPS into man (and correspondingly smaller amounts in experimental animals) results, within 4 hours, in the development of transient fever, with body temperatures reaching up to 40°C (104°F). Other effects include changes in the white cell count, enhancement of hormonal and enzymic activities (for instance proteolysis in various tissues and organs), stimulation of phagocytosis and other defence mechanisms of the body. Larger doses cause hemorrhages, tissue destruction, shock and death. The LPS are largely responsible for the clinical symptoms of such diseases as dysentery, typhoid fever and tularemia, as well as many cases of food poisoning.

Since these toxins are part of the bacterial cell, they have been named "endotoxins", to distinguish them from the exotoxins, such as those of the diphtheria, cholera, tetanus and botulism, that are secreted from the cells into the medium. The latter are all proteins, which are heat labile, in contrast to the endotoxins which are heat stable. In fact, many of the pharmacological effects of living Gram-negative bacteria, or of purified LPS, are also produced by injection of heat-killed cells or of heated LPS.

Before concluding this brief description of the biological properties of LPS, I would like to mention the recent finding that these compounds exert mitogenic effects on bone marrow derived (B)-lymphocytes, the main cells that produce and secrete immunoglobulins. LPS thus differ from most mitogenic lectins, such as concanavalin A or soybean agglutinin, which are specific for thymus derived (T)-lymphocytes. Of particular significance is the fact that stimulation of B-lymphocytes by LPS causes not only an increase in DNA and protein synthesis, and cell proliferation, but also leads to the secretion of immunoglobulins. LPS is therefore being used in following the biochemical changes that occur in B-lymphocytes after stimulation, and in particular of immunoglobulin biosynthesis and secretion (A2, A3).

As a result of the classic work done in the 1930's, mainly in the laboratories of A. Boivin in France and W. T. J. Morgan in London, it became clear that the complete somatic or O-antigen specificity of the microorganisms resides in the polysaccharide portion of LPS, which can be liberated from the complex by mild acid hydrolysis (0.1 M acetic acid, 100°C, 2 h). The isolated polysaccharide is non-toxic. The lipid moiety of the LPS, lipid A, is a complex phosphorylated lipid of unusual composition, which contains glucosamine instead of glycerol and is rich in 3-hydroxymyristic acid. The lipid appears to account entirely for the toxicity of the original endotoxin, as well as for its mitogenic activity (A4).

The physicochemical properties of LPS are determined by the simultaneous presence of hydrophilic, strongly polar polysaccharide regions, on the one hand, and of lipid A on the other. This combination of hydrophilic and hydrophobic regions results in LPS being an amphipathic material which in aqueous solutions forms micellar structures with a high degree of aggregation, thus making it impossible to determine its true molecular weight.

Most of the work in this field has been done on the LPS isolated from the enteric bacteria, the enterobacteriaceae, chiefly the *Salmonellae* and *E. coli*. Organisms of this class normally inhabit our intestinal tract, and are also the major causes of gastrointestinal diseases and of food poisoning. In the United States alone there are an estimated 2,000,000 human cases of *Salmonella* infections annually, making it one of the most important communicable diseases in the country. Another indication of the great interest in these organisms is the fact that today more than 1,000 serotypes (or species) of *Salmonella* are known, and new additions to this list become available each year. The separation of one serotype from another is based on differences in immunological specificities of the thermolabile flagellar H-antigens and the thermostable somatic O-antigens. If only the polysaccharide O-antigens are considered, the number of distinguishable antigenic types of *Salmonella* is over 60. Many of the *Salmonella* antigens cross-react with one another indicating that their polysaccharide determinants share common reactive groups. It is on the basis of these cross-reactions that F. Kauffmann in Denmark and B. White in England have classified the *Salmonellae* into major groups, designated by the letters A to I. More than 95% of the strains isolated fall into the first five groups (A to E).

This classification, the original purpose of which was to facilitate diagnosis of *Salmonella* infections, served as a basis for the structural and immunochemical studies of LPS in the laboratories of O. Westphal, A. M. Staub and P. W. Robbins. Their work has shown that the determinants for all O-antigen specificities of a given organism are located in a single polysaccharide molecule, and made it possible, in many cases, to identify the monosaccharide or oligosaccharide groupings which determine individual specificities.

The use of immunochemical reactions in structural polysaccharide chemistry is well established, and we have seen fine examples of results obtained with this approach in the study of the chemical structure of the ABH blood group substances. Such studies have also contributed greatly to the elucidation of the chemical structure of LPS. Another approach which was of particular importance in unravelling the structure of the bacterial LPS was the biosynthetic one, which I shall discuss in detail in the next lecture. It is based mainly on studies with bacterial mutants which lack the ability to synthesize certain intermediates necessary for the formation of the complete LPS molecule. These studies have demonstrated, among other things, that knowledge of structure is not always a prerequisite for the understanding of biosynthetic routes, and that structure may, in fact, be deduced from studies in biosynthesis.

COMPOSITION AND STRUCTURE

The LPS of *Salmonellae* have a highly complex chemical

composition. For example, LPS of *Salmonella typhimurium* is made up of some 15 constituents. It contains eight different sugars, phosphate, and ethanolamine in its polysaccharide portion, and various additional components in its lipid portion (Table 18).

Table 18

Constituents of LPS from *Salmonella typhimurium*

Polysaccharide	Lipid A
3-deoxy-mannooctulosonic acid (KDO)	glucosamine
	phosphate
L-glycero-mannoheptose	fatty acids
glucose	
galactose	
N-acetylglucosamine	
mannose	
L-rhamnose	
3,6-dideoxygalactose (abequose)	
phosphate	
ethanolamine	

You may note that in addition to common sugars, such as glucose, galactose, mannose and glucosamine, and the somewhat less common L-rhamnose (6-deoxy-L-mannose), we encounter here several unusual compounds:

$$
\begin{array}{ccc}
\text{COOH} & \text{CHO} & \text{CHO} \\
| & | & | \\
\text{C=O} & \text{HOCH} & \text{HCOH} \\
| & | & | \\
\text{CH}_2 & \text{HOCH} & \text{CH}_2 \\
| & | & | \\
\text{HOCH} & \text{HCOH} & \text{HOCH} \\
| & | & | \\
\text{HOCH} & \text{HCOH} & \text{HCOH} \\
| & | & | \\
\text{HCOH} & \text{HOCH} & \text{CH}_3 \\
| & | & \\
\text{HCOH} & \text{CH}_2\text{OH} & \\
| & & \\
\text{CH}_2\text{OH} & & \\
\end{array}
$$

2-Keto-3-deoxyoctanoic   L-glycero-   3,6-dideoxy-
acid or 3-deoxy-manno-   mannoheptose   galactose
octulosonic acid (KDO)   ("heptose")   (abequose)

Both 2-keto-3-deoxyoctanoic acid (or 3-deoxy-mannooctulosonic acid, KDO) and a heptose, usually L-glycero-mannoheptose, have been found in almost all LPS examined and appear to be confined to Gram-negative bacteria. KDO is an unusual sugar, first discovered in the LPS of *E. coli* by E. C. Heath and M. A. Ghalambor in 1963. It is similar to neuraminic acid in being a 2-keto-3-deoxy acid. In fact, it gives some of the colour reaction as the sialic acids - for example, it can be assayed by the periodate-thiobarbituric acid method, a general reaction specific for 2-keto-3-deoxy acids. The resemblance between KDO and the sialic acids is further borne out by the fact that the mechanisms of the biosynthesis of both sugars, as well as their mode of activation, are essentially the same.

Abequose, or 3,6-dideoxygalactose (3-deoxyfucose, 3,6-dideoxy-D-xylohexose) is a member of a new family of sugars, the 3,6-dideoxyhexoses, first discovered in *Salmonella* and *E. coli* strains. Four additional 3,6-dideoxyhexoses (of the 8 possible isomers) have been identified to date:

tyvelose	3,6-dideoxymannose
colitose	3,6-dideoxy-L-galactose
paratose	3,6-dideoxyglucose
ascarylose	3,6-dideoxy-L-mannose

The last member of the family, ascarylose, has not been found in microorganisms, but was isolated from the parasitic worm *Ascaris*.

When present in LPS, these dideoxyhexoses occur at the non-reducing ends of chains and constitute immunodominant groups for their respective antigenic determinants. Thus, paratose is a determinant of group A *Salmonella*, abequose of group B and tyvelose of group D. They are readily split off by mild acid hydrolysis, and since they migrate on paper chromatograms faster than any other sugar constituent of LPS, they can be easily detected and identified.

In addition to the 3,6-dideoxyhexoses, many other sugars have been isolated from the LPS of *Salmonella* and of other genera, most of which have not been found elsewhere. These include 2- and 3-amino-6-deoxyhexoses, for example 2-amino-2,6-dideoxyglucose(quinovosamine), 2-amino-2,6-dideoxy-D- and L-galactose (D- and L-fucosamine), and 2-amino-2,6-dideoxy-L-mannose (L-rhamnosamine). D-Fucosamine is a sugar we have been studying in our laboratory for a number of years; it was isolated by us in 1961 from the Gram-positive *Bacillus subtilis* and *Bacillus licheniformis*, and we also developed methods for its chemical synthesis. Of the dozen or so other unusual sugars found in different LPS, I shall only mention a few additional examples: 4-amino-4,6-dideoxy-D-glucose and D-galactose, and 2,3-diamino-2,3-dideoxy-D-glucose. So far, some twenty such unusual monosaccharides have been identified in LPS from various organisms, and the list is growing.

Obviously the structure of such complex macromolecules is extremely difficult to establish, though the fact that mild acid hydrolysis splits the LPS molecule into polysaccharide and lipid A gave an idea about the overall organization of LPS. The first clue concerning the structure of the polysaccharide came from the study of mutant strains in O. Westphal's laboratory in Germany. It was known that the wild type strains of enteric bacteria, which form smooth (S) colonies and have O-antigens, give rise to mutants forming rough (R) colonies and having none of the O-antigenic specificities. Westphal and his associates in the 1950's, analyzed the sugar composition of LPS from wild strains (S-forms) and from their mutants (R-forms) and found that LPS of all the R-forms examined contained only four sugars: glucose, galactose, heptose and glucosamine, whereas LPS of S-forms frequently contained additional sugars as already noted for the LPS of *S. typhimurium*. Moreover, it became clear that LPS of species belonging to the same serogroup in the Kauffmann-White scheme, contain identical sugar units. In other words, there exists a close relationship between the serotype of a species and the chemotype of its LPS.

From these results Westphal proposed the following hypothesis: (a) the polysaccharide is composed of two parts, an "R core" and "S-specific side chains" that are attached to the surface of the core (Fig.57); (b) the R core contains only the aforementioned four sugars, whereas S-specific side chains contain various other sugars; (c) biosynthesis of the polysaccharide begins from the R core; (d) R-forms can synthesize the R core normally but either cannot synthesize or cannot attach S-specific side chains, presumably owing to a defect in certain enzyme(s).

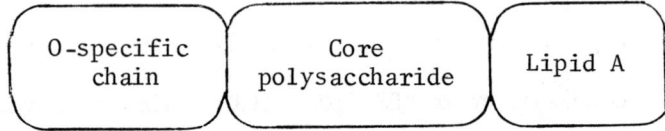

Figure 57   The three structural regions and their sequence in lipopolysaccharides

During the subsequent years all the predictions of this hypothesis turned out to be correct.

REFERENCES

Reviews

1. The bacterial cell wall,
    N. Sharon, Scient. American 220 (5), 92-98 (1969).
    A suitable introduction to the subject.

2. The Bacterial Cell Wall,
    M. R. J. Salton, Elsevier, 1964, 293 pp.
    The most comprehensive source on the early developments in the field, written by the man who laid the foundations to it.

3. Bacterial cell walls,
    D. J. Tipper, in Subunits in Biological Systems, Part B (Eds. G. D. Fasman and S. N. Timasheff), Biological Macromolecules Vol.6, Marcel Dekker Inc., New York, 1973, pp.121-205, 331-346.

4. Mode of action of antibiotics on microbial walls and membranes,
    M. R. J. Salton and A. Tomasz, Eds., Ann. N.Y.Acad. Sci. 235, 1-620 (1974).
    A collection of papers presented at a meeting of the

New York Academy of Sciences in June 1973, covering essentially all modern aspects of cell wall research.

5. Structure, biosynthesis and function of teichoic acids,
   J. Baddiley, Accoun. Chem. Res. $\underline{3}$, 98-105 (1970).

6. Teichoic acids in cell walls and membranes of bacteria,
   J. Baddiley, Essays Biochem. $\underline{8}$, 35-77 (1972).

7. The pneumococcal polysaccharides,
   M. J. How, J. S. Brimacombe and M. Stacey, Adv. Carb. Chem. $\underline{19}$, 303-358 (1964).

8. Biochemistry of bacterial cell envelopes,
   V. Braun and K. Hantke, Ann. Rev. Biochem. $\underline{43}$, 89-121 (1974).
   Contains a brief description of the properties and location in the cell wall of the lipoprotein of Gram-negative bacteria (pp.91-94).

9. Immunochemistry of O and R antigens of *Salmonella* and related *Enterobacteriaceae*,
   O. Lüderitz, A. M. Staub and O. Westphal, Bacteriol. Revs. $\underline{30}$, 192-255 (1966).
   The classical review of the subject.

10. Recent results on the biochemistry of the cell wall lipopolysaccharides of *Salmonella* bacteria,
    O. Lüderitz, Angew. Chem. Intern. Ed. $\underline{9}$, 649-663 (1970).

11. Microbial Toxins,
    G. Weinbaum, S. Kadis and S. J. Ajl, Eds., Bacterial Endotoxins, Vol.4, Academic Press, 1971, 473 pp.
    The most complete survey in which the chemical and physical structure of lipopolysaccharides, their biosynthesis and genetics, have been summarized by the top experts on these subjects.

Specific articles

A1. Pneumococcal C-substance, a ribitol teichoic acid containing choline phosphate,
    D. E. Brundish and J. Baddiley, Biochem. J. $\underline{110}$, 573-582 (1968).

A2. Mitogens as probes for immunocyte activation and cellular cooperation,
J. Andersson, O. Sjöberg and G. Möller, Transplant. Rev. 11, 131-177 (1972).

A3. Synthesis, surface deposition and secretion of immunoglobulin M in bone marrow-derived lymphocytes before and after mitogenic stimulation,
F. Melchers and J. Andersson, Transplant. Rev. 14, 76-130 (1973).

A4. The mitogenic effect of lipopolysaccharide on bone marrow-derived mouse lymphocytes. Lipid A as the mitogenic part of the molecule,
J. Andersson, F. Melchers, C. Galanos and O. Luderitz, J. Expt. Med. 137, 943-953 (1973).

18

LIPOPOLYSACCHARIDES - I: STRUCTURE

AND BIOSYNTHESIS

Although by the end of the 1950's the overall constitution of LPS was established, no structural details were available. Moreover, there was no evidence about the mechanism of LPS biosynthesis. A single LPS was believed to contain many different kinds of O-side chains, and some investigators felt that such complex structures could perhaps not be synthesized without some kind of template mechanism. The diversity of nucleotide "handles" in the various sugar nucleotides was thought to be useful for "coding".

USE OF DEFECTIVE MUTANTS

Elucidation of many details of LPS structure, as well as of the mechanism of its biosynthesis, came as a result of the use of bacterial mutants with known defects, first introduced into this field by H. Nikaido and T. Fukasawa in Japan about 1960 (1,2). These workers made the important observation that certain galactose negative mutants of *Salmonella* and *E. coli*, when grown on glucose, formed incomplete LPS in which not only galactose was missing, but also certain other sugars which are present in the wild type of LPS. The polysaccharide moiety of the deficient LPS was completely devoid of both O-specific side chains and the galactose and N-acetylglucosamine of the R core. The core did, however, contain glucose. The defect in LPS synthesis was shown to be the result of the absence of UDP-Gal-4-epimerase, an enzyme essential for the formation of UDP-Gal from carbon sources other than galactose (e.g. glucose) in the absence of galactose (Fig.58). To explain these findings, Nikaido postulated that completion of the LPS cannot take place without the attachment of the galactose to the core region, and that all the sugars missing in the mutant LPS must be located distal to this galactose moiety in the wild type (S-form) of LPS.

Indeed, when the mutants were grown for a short period in the presence of galactose, they synthesized an LPS indistinguishable in qualitative composition from the LPS of S-forms. Also, crude cell wall preparations (cell

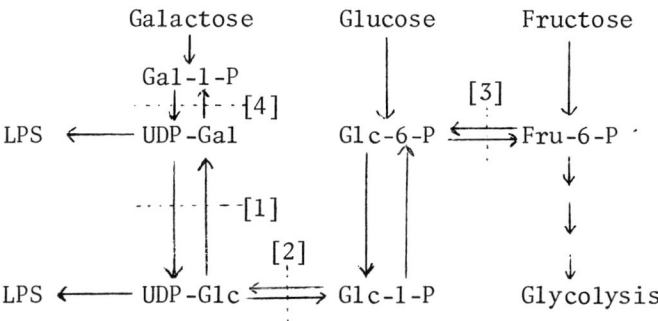

Figure 58 Pathways of the biosynthesis of UDP-Glc and UDP-Gal in enteric bacteria. The same pathways operate in higher organisms. By the reactions depicted, UDP-Gal can be synthesized either from exogenous galactose, or from exogenous glucose. Bacterial mutants are known which are deficient in enzymes catalyzing reactions [1], [2] or [3]. Reaction [1] is catalyzed by UDP-Gal-4-epimerase, [2] by UDP-Glc-pyrophosphorylase, [3] by phosphoglucose isomerase, and [4] by Gal-1-P-uridyltransferase.

envelopes) of the mutants transferred [^{14}C]galactose from UDP-[^{14}C]Gal into the incomplete LPS. Furthermore, after the initial attachment of the galactose, these preparations were found to incorporate other sugars as well, when incubated with the respective sugar nucleotides.

A second group of mutants, deficient in the biosynthesis of UDP-Glc, forms an incomplete lipopolysaccharide with a polysaccharide moiety composed only of heptose and KDO. Such mutants may lack either UDP-Glc-pyrophosphorylase (Fig.58, enzyme [2]), or phosphoglucose isomerase

(Fig.58, enzyme [3]), and hence are unable to synthesize UDP-Glc from exogenous glucose or fructose. As expected, these and other mutants, under suitable conditions, accumulate in their cells various sugar nucleotides (e.g. TDP-Rha, CDP-Abe, and when grown on galactose very large amounts of UDP-Gal), whose utilization for LPS synthesis is prevented by the genetic block. The identification of these compounds, together with their use in enzymic reactions, has contributed greatly to the elucidation of both the biosynthesis and structure of LPS.

Table 19

Quantitative sugar composition of the polysaccharide portion of LPS from mutant strains of *Salmonella typhimurium*

Mutant	Molar ratios of sugars				
	GlcNAc	Gal	Glc	Hep	KDO
Rough (Ra)	1	2	2	2	3
UDP-Gal-deficient	0	0	1	2	3
UDP-Glc-deficient	0	0	0	2	3

Comparison of the sugar composition of the core polysaccharide of the two groups of mutants, and that of the common rough forms (designated as R mutants) of *S. typhimurium*

(Table 19), led to the prediction that the overall structure of the core region is as follows:

GlcNAc, Gal ⟶ Glc ⟶ Hep, KDO ⟶

Subsequently, other mutants were isolated which formed incomplete core structures. These include very rough mutants which contain only lipid A and KDO in their LPS.

In analogy to the chemotypes of the S-forms mentioned in the last lecture, the lipopolysaccharides of the mutants could be classified into R chemotypes, Ra to Re, with Ra having all the sugar constituents of the core, and Re being the most defective one. Recently, mutants defective in the synthesis of KDO have been isolated (Al).

The synthesis of KDO proceeds by the following reaction, catalysed by the enzyme KDO-8-phosphate synthetase:

$$\begin{array}{c}\text{CHO}\\\text{HO-C-H}\\\text{H-C-OH}\\\text{H-C-OH}\\\text{CH}_2\text{-O-P-O}^{\ominus}\\\text{O}^{\ominus}\end{array} + \begin{array}{c}\text{COOH}\\\text{C-O-P-O}^{\ominus}\\\text{CH}_2\end{array} \rightarrow \begin{array}{c}\text{COOH}\\\text{C=O}\\\text{CH}_2\\\text{HO-C-H}\\\text{HO-C-H}\\\text{H-C-OH}\\\text{H-C-OH}\\\text{CH}_2\text{-O-P-O}^{\ominus}\end{array} + HPO_3^{\ominus}$$

D-arabinose-5-phosphate   phosphoenol-pyruvate   KDO-8-phosphate

Note the similarity to the reaction (given on p.149) in which NANA is formed from N-acetylmannosamine-6-phosphate and phosphoenolpyruvate. In one of the mutants the enzyme responsible for the synthesis of KDO is altered. The altered enzyme has a 30-fold elevated K_m for $\underline{D}$-arabinose-5-phosphate and requires exogenous $\underline{D}$-arabinose-5-phosphate to maintain an adequate internal concentration of this substrate. In the absence of arabinose-5-phosphate, the cells cease to grow after one generation but remain viable. It seems that the ultimate defect in the LPS structure permitting the cells to remain viable is reached with KDO mutants, and that the complete KDO-lipid A structure is essential for growth.

On the basis of extensive studies with these and other mutants, as well as chemical and enzymic degradation experiments, and recently of methylation analysis (A2), the detailed structure of the LPS has been formulated (Fig.59).

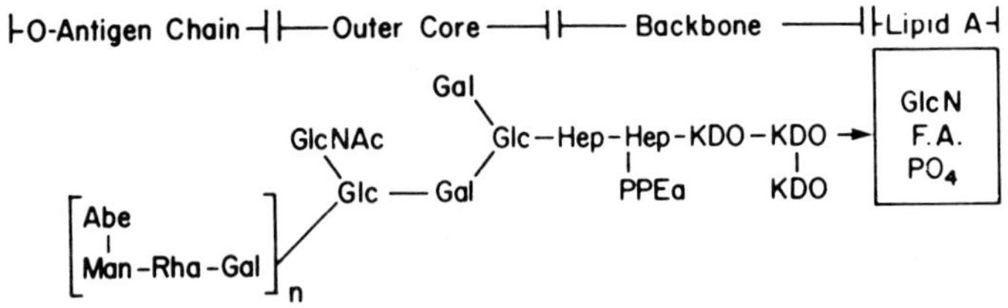

Figure 59 Structure of the lipopolysaccharide of *S. typhimurium*. Abe, abequose (3,6-dideoxygalactose); Ea, ethanolamine; F.A., fatty acid; P, phosphate. A number of such chains are crosslinked by phosphate bridges between the heptose residues, and by pyrophosphate bridges between the glucosamine residues of lipid A.

The structural investigations filled in many details that could not be obtained by the genetic and biosynthesis studies. These included the nature of the linkages between the sugars and the position of the different substituents (e.g. the phosphate linked to the heptose). For example, isolation of melibiose (Gal-α-(1→6)-Glc) from partial acid hydrolysates of Rb (formerly known as RII) mutants of *Salmonella*, provided evidence for the occurrence of a 1→6 linked galactose branch on the linear polysaccharide chain of the R-core. A number of other oligosaccharides have been isolated from Ra (RI) and Rb mutants, including the pentasaccharide

```
                    Gal
                     ↓
GlcNAc → Glc → Gal → Glc
```

Partial acid hydrolysis of the LPS from Rc and Rd mutants after removal of the phosphate, or from mutants in which no phosphate is linked to the heptose (P$^-$ mutants), afforded the oligosaccharides Glc-Hep-Hep-KDO and Hep-Hep-KDO. Periodate oxidation of the trisaccharide under conditions which led to almost exclusive cleavage between the C-6 and C-7 of the heptose, followed by borohydride treatment, degraded the heptose into mannose (Fig.60). The resultant product was digested by α-mannosidase, establishing that the linkage of the heptose residues is of the α-configuration (A3).

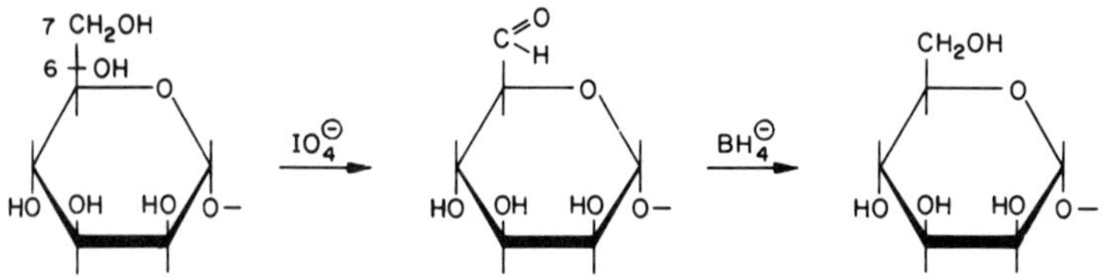

Figure 60 Conversion of the heptose into mannose by controlled periodate oxidation and borohydride reduction.

STRUCTURE OF THE O-SPECIFIC CHAINS (3)

Our knowledge of the structure of this region, which is completely different from that of the R-core, came originally from the work of P. W. Robbins and his colleagues at M.I.T. in Boston, and was based to a large extent on the application of immunochemical techniques. They isolated di-, tri- and tetrasaccharides from partial acid hydrolysates of the LPS of *Salmonella anatum* and from related strains of group E *Salmonella*, and determined the ability of these oligosaccharides to inhibit the interaction of the parent LPS with its antibody. In addition they established the structure of the di- and trisaccharides by methylation, periodate oxidation and with the aid of specific glycosidases. From the results obtained they concluded that the O-side chain in this organism is made up by the polymerization of a single trisaccharide repeating unit, Man-Rha-Gal.

I would like to emphasize that the concept that the O-side chains are long, was entirely new. Before this work people were inclined to think that the O-antigen had a huge core, hard to define structurally, and very short side chains, containing only a few monosaccharide units. The new concept was subsequently found to be valid in many other organisms and, wherever careful studies were made, O-side chains were found to consist of oligosaccharide repeating units, usually tri- or tetrasaccharides, but sometimes as big as a hexasaccharide. Thus, A. M. Staub and her co-workers at the Pasteur Institute in Paris found that in the oligosaccharide side chain of the LPS from group B *Salmonella*, such as *S. typhimurium*, the repeating unit is similar to that found in group E, except that both glucose and abequose are attached to it (Fig.61). The precise number of repeating trisaccharide units per side chain is not known for any organism, but is believed to be between 6 and 15.

The above examples demonstrate how the immunological specificities of O-antigens are determined by relatively minor structural differences in the side chains of otherwise identical polysaccharides. The modifications that have been observed in closely related antigens include: [1] changes in position of linkages, i.e. 1→4 *vs* 1→6; [2] altered anomeric

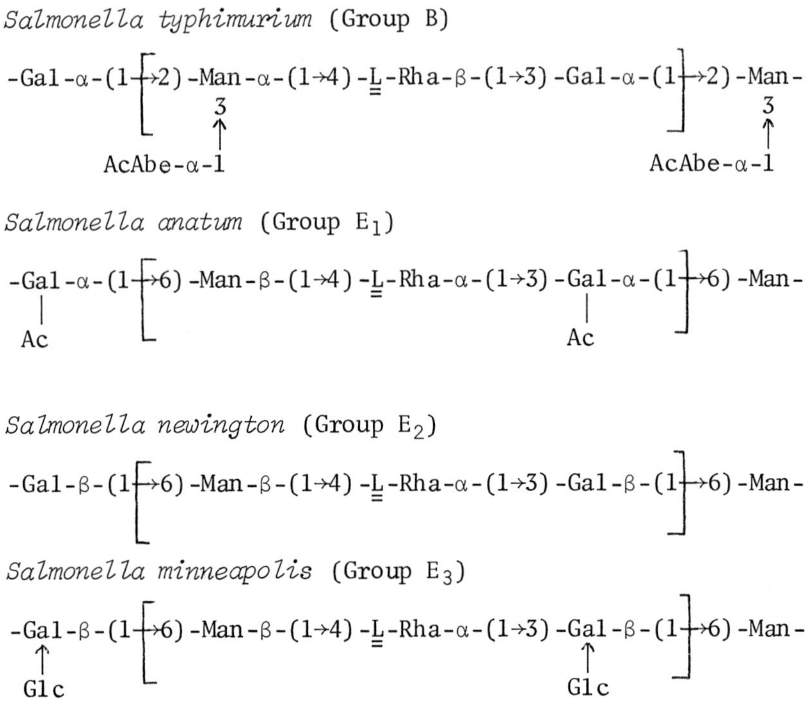

Salmonella typhimurium (Group B)

$$-Gal-\alpha-(1\rightarrow 2)-Man-\alpha-(1\rightarrow 4)-\underline{L}-Rha-\beta-(1\rightarrow 3)-Gal-\alpha-(1\rightarrow 2)-Man-$$
$$\underset{AcAbe-\alpha-1}{\overset{3}{\uparrow}} \qquad\qquad \underset{AcAbe-\alpha-1}{\overset{3}{\uparrow}}$$

Salmonella anatum (Group E$_1$)

$$-Gal-\alpha-(1\rightarrow 6)-Man-\beta-(1\rightarrow 4)-\underline{L}-Rha-\alpha-(1\rightarrow 3)-Gal-\alpha-(1\rightarrow 6)-Man-$$
$$\underset{Ac}{|} \qquad\qquad \underset{Ac}{|}$$

Salmonella newington (Group E$_2$)

$$-Gal-\beta-(1\rightarrow 6)-Man-\beta-(1\rightarrow 4)-\underline{L}-Rha-\alpha-(1\rightarrow 3)-Gal-\beta-(1\rightarrow 6)-Man-$$

Salmonella minneapolis (Group E$_3$)

$$-Gal-\beta-(1\rightarrow 6)-Man-\beta-(1\rightarrow 4)-\underline{L}-Rha-\alpha-(1\rightarrow 3)-Gal-\beta-(1\rightarrow 6)-Man-$$
$$\underset{Glc}{\uparrow} \qquad\qquad \underset{Glc}{\uparrow}$$

Figure 61 Antigen repeating sequences in the B and E groups of *Salmonella*. Note that these different antigens represent variations of the same basic repeating sequence, Man-Rha-Gal. The variations are in the linkage positions, their anomery, as well as in substitution on the sugars of the main chain. Ac - acetyl.

configuration (α or β); [3] attachment of additional monosaccharides, such as glucose, at different positions on the repeating sequence; [4] deletions of, or substitution for, one of the monosaccharides in the basic trisaccharide unit; and [5] the presence or absence of acetyl groups on one type of sugar residue.

LIPID A

Lipid A is a common constituent of all S and R forms of LPS. No mutants have been identified so far which are defective in its biosynthesis - presumably because such a block may be lethal to the cell. This is probably one of the main reasons why nothing is known of the mechanism of the biosynthesis of lipid A.

Structural studies of lipid A have been performed on preparations isolated from partial acid hydrolysates of LPS. Such preparations are heterogenous and are degraded to varying extents. Because of this the structure of lipid A has not yet been fully established. Work on the subject has been done mainly by O. Westphal, O. Lüderitz and their colleagues in Freiburg (A4), who have shown that lipid A is composed of a disaccharide of glucosamine, linked β-(1→6), to which phosphate groups are attached (Fig.62). Apparently the phosphate forms a diester (or pyrophosphate) bridge to the C-4 of the non-reducing terminal of the glucosaminyl residue of a neighbouring chain, thus serving as a crosslink between LPS chains. All available hydroxyl and amino groups of the GlcN-β-(1→6)-GlcN disaccharide are acylated, chiefly by a 14 carbon fatty acid, 3-hydroxymyristic acid, found exclusively in lipid A.

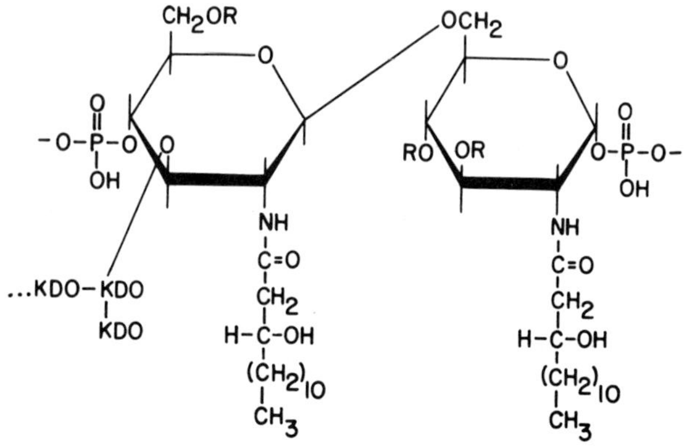

Figure 62 Proposed structure of the KDO-lipid A region of LPS (A4). R, different long chain fatty acids, e.g. lauric, myristic and palmitic. The amino groups are substituted by 3-hydroxymyristic acid.

BIOSYNTHESIS OF THE CORE POLYSACCHARIDE

The core polysaccharide (outer core and backbone in Fig.59), appears to be uniform in all *Salmonellae*. A somewhat different core is present in *E. coli* and in other major groups of the enteric bacteria.

The biosynthesis of the core presumably begins with the transfer of KDO from cytidine monophospho-KDO (CMP-KDO) (Fig.63) to lipid A. Here we see another point of similarity between KDO and the sialic acids - the activated form of both is a nucleoside monophosphate and not diphosphate (e.g. uridine diphosphate glucose), as is generally found. In fact, these are the only two nucleoside monophosphate sugars that have been recognized to date.

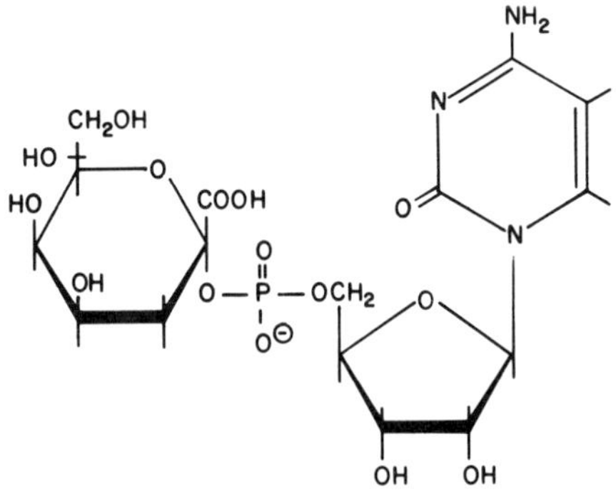

Figure 63 Structure of CMP-KDO. The linkage between the phosphate and the anomeric carbon (C-2) of KDO is depicted here as being in the α-configuration, in common with other sugar nucleotides, but no proof has been presented for this.

Transfer of KDO to lipid A was demonstrated in a cell free system from *E. coli* 0111 but not in *Salmonella*. The enzymically transferred KDO appears to be glycosidically linked, since the potential carbonyl group at C-2 of KDO could not be reduced by sodium borohydride. The site of attachment of KDO in the acceptor molecule is not yet known; presumably it is one of the hydroxyl groups of the disaccharide of lipid A, or that of the 3-hydroxymyristoyl group. Little is known regarding the biosynthesis of the remaining portion of the inner region of the core polysaccharide. There is evidence for the presence of a branched trisaccharide of KDO, at least in *Salmonella* LPS (Fig.62), but no data are available on the enzymic addition of the second and third KDO residues.

One, or both, of the heptose residues in LPS are phosphorylated, and different polysaccharide chains may be crosslinked by phosphodiester linkages. Some mutants of *Salmonella* (P^- mutants) produce LPS lacking these phosphate groups, and the LPS in their cell envelope fraction acts as an acceptor of the terminal phosphate group of ATP in a cell-free system according to the sequence shown below:

$$(Hep)_2\text{-}(KDO)_3\text{-Lipid A} \xrightarrow{UDP\text{-}Glc} Glc\text{-}(Hep)_2\text{-}(KDO)_3\text{-Lipid A} \xrightarrow{ATP} \underset{\underset{P}{\uparrow}}{Glc\text{-}(Hep)_2\text{-}(KDO)_3\text{-Lipid A}}$$

It is of interest that both in intact cells and in a cell-free system, LPS lacking these phosphate groups can accept the first glucosyl group of the outer core, but not the next sugar, namely Gal. The phosphorylating enzyme appears to prefer LPS containing the heptose substituted by glucose to that devoid of the latter sugar and phosphorylation normally takes place after the addition of the first glucose but before the addition of the other monosaccharides of the outer core (A5).

In contrast to the limited information about the synthesis of the inner core (or backbone), there is now ample evidence on the biosynthesis of the outer core, mainly due to the efforts of B. L. Horecker, M. J. Osborn and L. I. Rothfield at the Albert Einstein College of Medicine in New York. The outer core is formed by the successive incorporation of glucose, galactose, glucose and N-acetylglucosamine from their corresponding UDP-derivatives. The reactions were initially studied in crude cell wall preparations (envelope fractions) of suitable mutants (Fig.64).

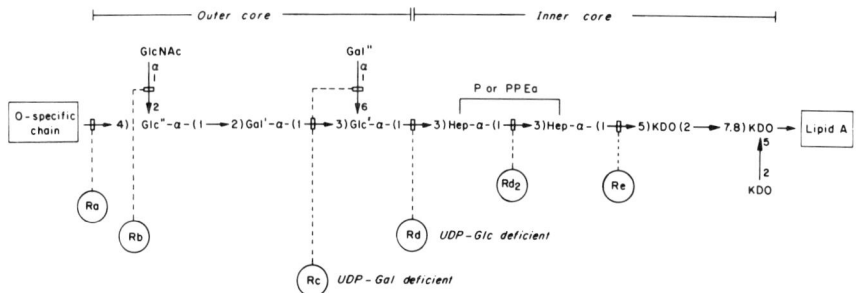

Figure 64 Structure of the core region of the lipopolysaccharide of *S. typhimurium*. The dotted lines indicate the structures of some of the different R mutants (Ra - Re). Hep, L-glycero-D-mannoheptose; Ea, ethanolamine. A number of chains are crosslinked by phosphate bridges between the heptose residues, and by pyrophosphate bridges between the glucosamine residues of lipid A.

The incorporation of each of the sugars is specifically dependent on the presence of the preceding sugar in the LPS. The first reaction is the transfer of glucose (Glc') to an LPS obtained from Rd mutants which contains only heptose and KDO in its core. The enzyme catalyzing this reaction is glucosyltransferase I. The enzymic transfer of the next sugar residue, that of galactose (Gal' in Fig. 64) was accomplished by using the cell envelope preparation from mutants defective in UDP-Gal synthesis. This reaction is catalyzed by galactosyltransferase I. The site of attachment of the transferred galactose was difficult to determine, because the galactosyl linkage formed was extremely acid-labile. This problem was finally solved by B. L. Horecker and his coworkers in a rather ingenious way (Fig.65). The primary alcohol group at C-6 of the transferred galactosyl group was oxidized into an aldehyde group by the action of galactose oxidase. The aldehyde was further oxidized by hypobromite to a carboxyl. Since the glycosidic linkages of uronic acids are resistant to acid hydrolysis (p. 62), mild treatment of the oxidized LPS with acid produced an α-(1→3) disaccharide of galacturonic acid and glucose (GalUA-α-(1→3)-Glc). This result clearly showed that the galactose is linked to the glucose residue of the acceptor LPS through an α-(1→3) linkage.

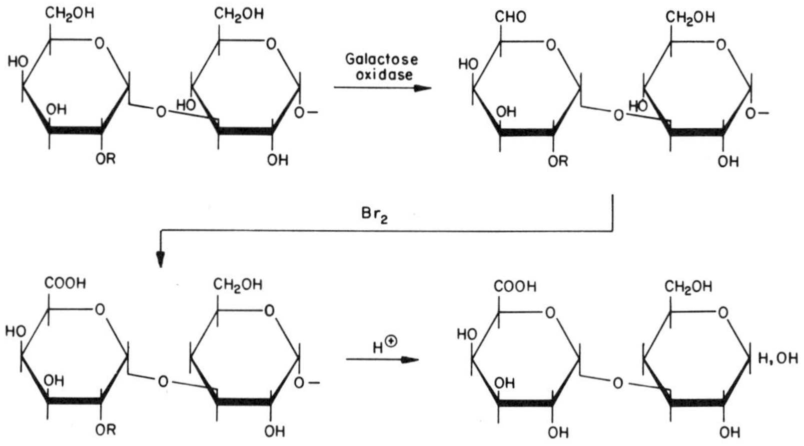

Figure 65 Sequence of reactions used for the identification of the Gal-α-(1→3)-Glc linkage in the core region of LPS.

In the early stages of the study on core synthesis, cell envelope preparations served as the source of both acceptor and enzyme; for this reason detailed work on the properties of the glycosyltransferases was difficult. The situation was, however, greatly improved when LPS in heated cell envelope preparations was found to serve as an exogenous acceptor. By the use of this acceptor, enzyme activities were also detected in the soluble fraction of sonic extracts of *S. typhimurium*, and it was established that the glucosyl- and galactosyltransferases have strict acceptor specificities, so that each of them only transfers glycosyl groups to the glucose or galactose-deficient LPS, respectively. This important finding established that the sequence of various sugars in the core is determined by the donor and acceptor specificities of glycosyltransferases which act in succession.

Both the particulate and soluble galactosyltransferases of *S. typhimurium* appear to catalyze the transfer of galactose only to the 3 position of glucose. From structural studies, however, it was known that the core polysaccharide contains a second galactose residue (Gal" in Fig.63) linked as an α-(1→6) branch to the proximal glucose residue (Glc') attached to the heptose. Galactosyltransferase II, the enzyme responsible for the addition of this branch galactosyl residue, has not been identified directly, but evidence for its existence has been obtained by analysis of mutants deficient in galactosyltransferase I.

The final reactions in biosynthesis of the core have been characterized in cell envelopes of mutants of *S. typhimurium*. They involve the transfer of the penultimate glucose residue (Glc") to the galactose (Gal') of the core, catalyzed by glucosyltransferase II, an enzyme different from glucosyltransferase I, and the attachment of N-acetylglucosamine to Glc", catalyzed by an N-acetylglucosaminyltransferase. Neither enzyme has been purified, nor is much known about their properties. This is in contrast to the situation with glucosyltransferase I and galactosyltransferase I which have

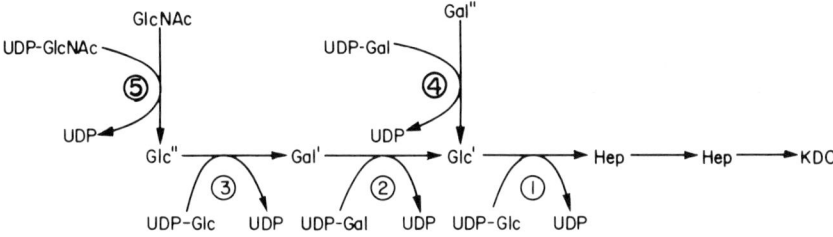

Figure 66   Enzymic reactions in the biosynthesis of the outer core region of the lipopolysaccharide of *S. typhimurium*. Enzymes: [1] glucosyltransferase I; [2] galactosyltransferase I; [3] glucosyltransferase II; [4] galactosyltransferase II; [5] N-acetylglucosaminyltransferase.

been purified and studied in detail.

The enzymic steps leading to the synthesis of the outer core are summarized in Fig. 66.

ROLE OF PHOSPHOLIPIDS IN SUGAR TRANSFER (4)

The availability of purified glucosyltransferase I and galactosyltransferase I, permitted a detailed study of the mode of action of these enzymes, and led to the discovery of the role of membrane phospholipids in this enzyme system. In 1964, L. I. Rothfield and B. L. Horecker made the important observation that although the heated cell envelope fraction was a good acceptor of glycosyl groups, purified LPS was completely inactive. This paradox was solved by them when they found that the removal of phospholipids from cell envelope preparations led to the loss of their acceptor activity. Phospholipids could be added back to the extracted cell envelope fraction, or even to the purified LPS, and heating and slow cooling of the mixture was found to produce a phospholipid-LPS or phospholipid-cell envelope complex, each of which was fully active as an acceptor of glycosyl groups. Electron microscope studies showed that phospholipid molecules were inserted between LPS molecules, and that together these two kinds of molecules formed mixed micelles in the aqueous environment. This system is an interesting example in which phospholipids are required in a glycosyltransferase reaction.

Rothfield has further found that the LPS-phospholipid complexes appear to be the true substrates of the transferases. In addition to LPS-phospholipid, LPS-phospholipid-enzyme complexes were also isolated by centrifugation through sucrose density-gradients. The binding of the transferases in this system is very specific. LPS alone does not bind any enzyme; LPS preparations of different structure, when complexed with phospholipids, bind only the transferase for which these preparations can serve as substrates. Thus, the complex of glucose-deficient LPS and phospholipids binds only glucosyltransferase I and the complex containing galactose-deficient LPS binds only galactosyltransferase I. These results again indicate the strict specificity with which the glycosyltransferases interact with acceptors, and suggest an obligatory reaction sequence of the following type:

$$LPS + PEa \longrightarrow LPS.PEa$$

$$Enz + LPS.PEa \xrightarrow{Mg^{++}} Enz.LPS.PEa$$

$$UDP\text{-}Gal + Enz.LPS.PEa \xrightarrow{Mg^{++}} Gal\text{-}LPS.PEa + UDP + Enz$$

where LPS is a galactose-deficient lipopolysaccharide, PEa is phosphatidylethanolamine, and Enz is galactosyltransferase I. A similar reaction sequence holds for the transfer of glucose from UDP-Glc by glucosyltransferase I to a suitable acceptor.

Reconstitution of the functional enzyme system in a monolayer by stepwise formation of the binary LPS.PEa complex and ternary Enz.LPS.PEa from an initial monomolecular film of phosphatidylethanolamine suggests that the arrangement of molecules in the monolayer mimics that in the native cell envelope. The detailed mechanism by which phospholipids produce active substrates out of LPS is not completely clear. The observation that phospholipid molecules become inserted between LPS molecules suggests that phospholipids might function simply by "supporting" and holding LPS molecules in a particular conformation, so that LPS could interact with the active site of the enzyme. A model, proposed by Rothfield, which incorporates these ideas, is shown in Fig. 67.

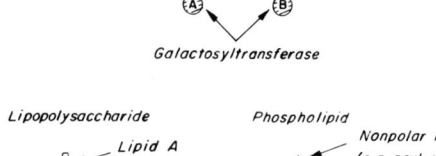

Figure 67 A speculative model of the membrane containing the galactosyltransferase enzyme system. Two possible locations of the enzyme are shown.
A. A portion of the enzyme penetrates into the nonpolar interior of the membrane. B. Enzyme is located only in the polar portion of the membrane. Present evidence does not permit a choice between the two possibilities (4).

Such reconstituted functional enzyme systems offer the possibility to study the sequential growth of the LPS, and the movement of the substrates and/or the enzyme within the membrane. Studies of this kind are of particular interest since they may provide clues to enzyme-directed translocation processes in membranes.

REFERENCES

Reviews

1. Biosynthesis of cell wall lipopolysaccharide in Gram-negative enteric bacteria,
    H. Nikaido, Advan. Enzymol. 31, 77-124 (1968).
    A most basic article on the subject. Deals also with structure of LPS, the study of which went hand in hand with the study of biosynthesis. Highly recommended.

2. *See Lecture 17, references 9, 10 and 11.*

3. Determinants of specificity in *Salmonella*: changes in antigenic structure mediated by bacteriophages,
    P. W. Robbins and T. Uchida, Fed. Proc. 21, 702-710 (1962).
    A very well written, early review of the classical work on the subject.

4. Enzyme reactions in biological membranes,
    L. I. Rothfield and D. Romeo *in* Structure and Function of Biological Membranes (Ed. L. I. Rothfield), Academic Press, 1971, pp.251-284.

Specific articles

A1. Isolation of a mutant of *Salmonella typhimurium* dependent on D-arabinose-5-phosphate for growth and synthesis of 3-deoxy-D-mannooctulosonate (ketodeoxyoctonate),
P. D. Rick and M. J. Osborn, Proc. Nat. Acad. Sci. USA 69, 3756-3760 (1972).

A2. Structural studies of the common-core polysaccharide of the cell-wall lipopolysaccharide from *Salmonella*,
C. G. Hellerqvist and A. A. Lindberg, Carbohyd. Res. 16, 39-48 (1971).
One of a series of papers dealing with methylation analysis of LPS. For a similar study of the O-specific chains, see Carbohyd. Res. 16, 297-302 (1971).

A3. Biochemical studies on lipopolysaccharides of *Salmonella* R mutants. 3. The linkage of the heptose units,
W. Dröge, O. Lüderitz and O. Westphal, Europ. J. Biochem. 4, 126-133 (1968).

A4. Nature and linkages of the fatty acids present in the lipid A component of *Salmonella* lipopolysaccharide,
E. Th. Rietschel, H. Gottert, O. Lüderitz, and O. Westphal, Europ. J. Biochem. 28, 166-173 (1972).

A5. Biosynthesis of *Salmonella* lipopolysaccharide. Studies on the transfer of glucose, galactose and phosphate to the core in a cell free system,
P. F. Mühlradt, Europ. J. Biochem. 18, 20-27 (1971).

19

LIPOPOLYSACCHARIDES - II: BIOSYNTHESIS

BIOSYNTHESIS OF O-SPECIFIC CHAINS

The synthesis of the O-specific side chains (or O-antigens) proceeds by an unusual mechanism, which differs in many respects from that of the synthesis of the core polysaccharide (1-4). The oligosaccharide repeating units of the O-side chains are first assembled on a lipid carrier, and only then is the completed chain joined to the core portion of the R-lipopolysaccharide. Although the different sugars of the outer core are all donated by their UDP derivatives, this is not the case for the synthesis of the O-specific chains. For the formation of the latter, galactose is transferred from UDP-Gal, $\underline{L}$-rhamnose comes from TDP-$\underline{L}$-Rha, mannose from GDP-Man, and the dideoxy sugars when present from their CDP derivatives (e.g. CDP-Abe).

The discovery that a lipid carrier is required for the synthesis of the O-specific chains was made by P. W. Robbins and his coworkers in 1965, at the same time as J. L. Strominger and his colleagues found that such a carrier is involved in the biosynthesis of peptidoglycan. Because the lipid carrier in bacteria is present in very small amounts, and is labile to both acid and alkali, its structure was difficult to establish. However, application of mass spectrometry solved the problem, indicating clearly that the lipid is a phosphate ester of a C_{55} polyisoprenoid alcohol of the following structure:

$$^{\ominus}O-\overset{O}{\underset{\underset{\ominus O}{|}}{\overset{\|}{P}}}-O-CH_2-CH=\overset{CH_3}{\underset{|}{C}}-CH_2-(CH_2-CH=\overset{CH_3}{\underset{|}{C}}-CH_2)_9-CH_2-CH=\overset{CH_3}{\underset{|}{C}}-CH_3$$

This lipid has been called by a variety of names: antigen-carrier lipid (ACL) phosphate, glycosyl carrier lipid (GCL) phosphate, undecaprenyl phosphate or bactoprenol phosphate.

Polyisoprenoid compounds are widely distributed in nature, but their function, particularly in bacteria, remained completely unknown until their identification in these pioneering studies as membrane-bound carrier lipids involved in the biosynthesis of the O-side chains of LPS and peptidoglycan (p.427). The same lipid is now known to act as the intermediate carrier of glycosyl units in the biosynthesis of several other bacterial polysaccharides (5) such as the capsular polysaccharide of *Aerobacter aerogenes*, of the mannan associated with the membrane of *M. luteus* and perhaps in bacterial lipoteichoic acid. Similar lipids (dolichol phosphates) seem to be involved in the biosynthesis of animal glycoproteins (p.169).

In bacteria, the lipid intermediates are located in the cell membrane and presumably serve to transport the activated sugars (in the form of their nucleotide derivatives) from the site of their synthesis inside the cell to the site of their utilization for the synthesis of "extracellular" complex polysaccharides.

The mechanism of biosynthesis of O-antigen has been most extensively studied in *S. typhimurium* in which the basic repeating unit is the branched tetrasaccharide

Abe
|
-Man-L-Rha-Gal and in *S. newington* and related species, which contain a similar repeating unit, the trisaccharide Man-L-Rha-Gal (Fig.61). The pathway of biosynthesis which emerged from these studies is summarized in Fig. 68. The entire reaction sequence is catalyzed by the cell envelope fraction which contains both the GCL and the final lipopolysaccharide

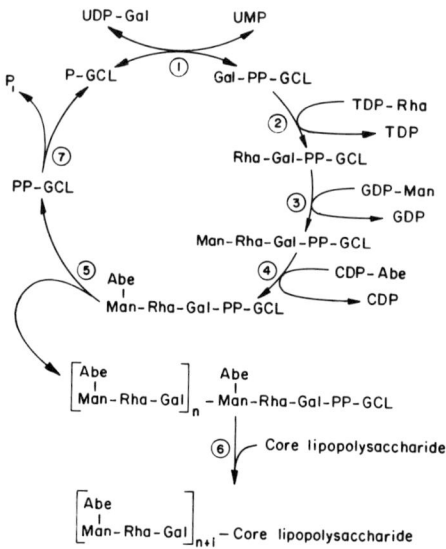

Figure 68   Pathway of biosynthesis of the O-specific side chain of *S. typhimurium*.

acceptor of the O-antigen chains together with all necessary enzyme proteins. The pathway can be divided into three main steps:

(a) Assembly of the oligosaccharide repeating unit as a lipid linked oligosaccharide (usually tri- or tetrasaccharide) intermediate (reactions 1-4, Fig.68).

(b) Polymerization of the repeating units to form the O-specific polysaccharide chain still linked to the carrier lipid (reaction 5).

(c) Transfer of the completed polysaccharide chain from the

carrier lipid to the lipopolysaccharide core (reaction 6).

The cycle is completed by regeneration of the lipid carrier phosphate in a reaction which removes the terminal phosphate from the lipid pyrophosphate (PP-GCL) (reaction 7). This pyrophosphatase reaction is sensitive to bacitracin, an antibiotic which also inhibits peptidoglycan biosynthesis.

The initial step in the sequence (reaction 1) is the transfer of galactose 1-phosphate from UDP-Gal to the lipid carrier phosphate P-GCL, to form the monosaccharide lipid intermediate, with stoichiometric release of UMP. This reaction is unusual since the sugar is transferred together with the phosphate moiety, and a nucleoside monophosphate (UMP), not a diphosphate (UDP), is released. Moreover, in contrast to other transfer reactions of sugar nucleotides that we have considered, Gal-PP-GCL is formed in a reversible reaction. Its equilibrium constant is approximately 0.5, indicating that the bond formed is of high energy. This high energy bond remains intact until the final transfer of sugar residues from the lipid intermediate to form a glycosidic bond with the acceptor polysaccharide (the R core).

In the presence of TDP-L-Rha, Gal-PP-GCL is efficiently converted to the disaccharide intermediate L-Rha-Gal-PP-GCL (reaction 2, Fig.68). This disaccharide derivative will then accept mannose from GDP-Man to form the trisaccharide intermediate Man-L-Rha-Gal-PP-GCL (reaction 3, Fig.68). With crude envelope preparation from *S. typhimurium*, the latter intermediate will accept abequose from CDP-Abe (reaction 4, Fig.68). In this way the complete repeating unit of the O-specific chain is obtained as an oligosaccharide-PP-GCL derivative. Polymerization of both the tri- and tetrasaccharide intermediates occurs very rapidly above 20°C, but formation and accumulation of the monomeric oligosaccharide-lipid intermediates (e.g. Man-L-Rha-Gal-PP-GCL) could be demonstrated by carrying out the incubations at 10°C.

Reactions 2, 3 and 4 (Fig.68) have not been characterized in detail. It is assumed that they are conventional glycosyltransfer reactions which yield the nucleoside diphosphate as products and are not readily reversible.

The three transferases responsible for synthesis of the trisaccharide-PP-GCL in *S. typhimurium* have been obtained in a soluble form by extraction of the cell envelope with non-ionic detergents. Evidence for a pyrophosphate linkage between the glycosyl unit and the lipid was obtained by brief hydrolysis of purified L-Rha-Gal-PP-GCL in 0.01 M HCl, which gave a quantitative yield of free disaccharide and inorganic pyrophosphate.

Chemical synthesis of Gal-PP-ficaprenol (a readily available geometric isomer of the natural glycosyl carrier lipid) has recently been accomplished by C. D. Warren and R. W. Jeanloz (A1). The synthetic product (Fig.69) could replace the natural one in the sequence of cell free reactions leading to the synthesis of the O-specific chains.

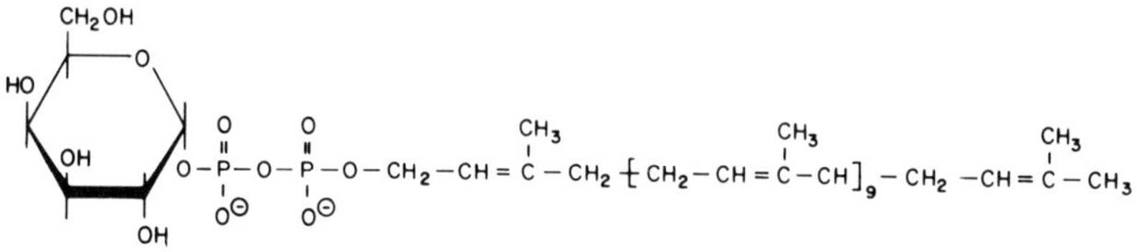

Figure 69 Structure of ficaprenol pyrophosphate galactose. The galactose is α-linked (A1).

MECHANISM OF POLYMERIZATION

As mentioned, when Man-L-Rha-Gal is produced, the trisaccharide units rapidly polymerize to form the O-specific chain polysaccharide. The occurrence of lipid-linked polymer chains was initially inferred in 1965 from analysis of the products of O-antigen synthesis in the cell envelope fraction of a mutant which produced an incomplete LPS core structure lacking the attachment site (Glc" in Fig.64) for the O-specific chains. The enzymic product was identified as a polysaccharide with the structure of O-specific side chains but which was not attached to a lipopolysaccharide. It is of interest that a similar polysaccharide, known as "haptenic O-antigen", was found to accumulate *in vivo* in a variety of core-defective mutants of *Salmonella*. This polysaccharide contained all the sugar components of the O-specific side chain, but lacked the core and lipid A, and was of lower molecular weight than LPS.

Direct evidence that the initial product of the enzymic polymerization is still attached to the carrier lipid was provided by the isolation of an intermediate, soluble in chloroform-methanol, which contained a dimer of the repeating unit,

$$\text{Man-}\underline{\text{L}}\text{-Rha-Gal-Man-}\underline{\text{L}}\text{-Rha-Gal-PP-GCL}$$
with Abe branches on each Man. Long chain polysaccharide-lipid derivatives (which are not soluble in organic solvents) have not been isolated in intact form due to degradation of the labile Gal-1-pyrophosphoryl-GCL linkage during their extraction from the incubation mixtures.

Abequose residues in *S. typhimurium* occur as branches of the O-side chain, and are not found in its main chain. The enzymic polymerization of the O-side chains readily occurs in the absence of CDP-Abe. Hence, it was thought possible that abequose residues are added after completion of the synthesis of the O-side chain. However, this hypothesis appears unlikely in view of the following findings. (a) Under certain experimental conditions, the incorporation of mannose, L-rhamnose, and galactose is strongly stimulated by the presence of CDP-Abe. (b) By skilful manipulation of the conditions of reaction, especially temperature, Man-L-Rha-Gal-PP-GCL was shown to be a good acceptor of abequose residues from CDP-Abe, whereas a cell envelope fraction containing the PP-GCL-linked polymer of this trisaccharide was inactive. Also, below 20°C the polymerization of repeating units was shown to be considerably faster in the presence of CDP-Abe than in its absence. (c) A *Salmonella* mutant that was defective in the biosynthesis of CDP-Abe was unable to polymerize the trisaccharide repeating units in intact cells. These results all favour the postulated sequence shown in Fig. 68, where abequose is added to the trisaccharide PP-GCL, and the resultant tetrasaccharide PP-GCL participates in the polymerization reaction.

The above mechanism of side chain modification at the trisaccharide-lipid intermediate does not appear, however, to be a general one. In two *Salmonella* strains in which glucose is attached to the galactose of the Man-L-Rha-Gal, modification of the carbohydrate chain was found to take place after, and not before, polymerization, while it is still linked to the lipid carrier (A2,A3). Moreover, in both strains, glucose is attached to the main O-specific chain by transfer from a glucosyl lipid intermediate, most likely Glc-P-GCL. Here phosphate is not transferred with glucose (see also p.388).

DIRECTION OF CHAIN GROWTH (6)

When an O-side chain is elongated, a monomeric repeating unit has to interact with an oligomer of repeating units. Both are linked to carrier molecules, most probably to PP-GCL. Here, two possible mechanisms, shown

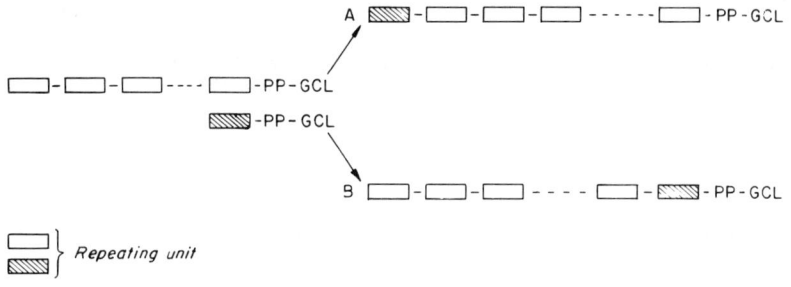

Figure 70   Two theoretically possible methods of chain elongation in O-side chain synthesis, as considered by Robbins et al. (6). In method A, repeating unit monomer is added to the non-reducing end of the growing O-side chain which is linked to PP-GCL. In method B, the repeating unit polymer is transferred onto the non-reducing terminal sugar of the repeating unit monomer, linked to PP-GCL.

in Fig. 70, may be envisaged, both of which proceed through the transfer of a substituted galactosyl residue onto a non-reducing terminal mannosyl residue of another repeating unit. But, whereas in one mechanism the O-side chain "grows" at the non-reducing end (A in Fig.70), in the other the addition of new repeating units takes place at the reducing end of the polysaccharide (B in Fig.70). The elegant pulse-chase experiments of Robbins *et al.*, using a mutant deficient in core synthesis, clearly showed that both *in vivo* and *in vitro* chain elongation is accomplished by the latter mechanism, that is by growth at the reducing end (Fig.71). This situation is rather unusual, because

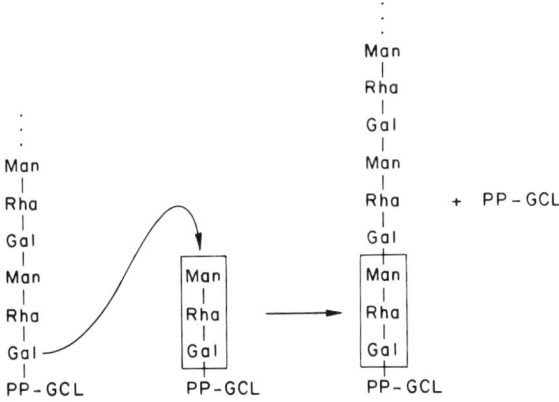

Figure 71 Mechanism of polymerization of O-antigen.

most polysaccharides are known to be synthesized by the transfer of a glycosyl group to the non-reducing end. It must, however, be emphasized that these examples of "growth at the non-reducing end" all involve direct transfer of a glycosyl group from a sugar nucleotide, unlike the elongation of O-side chains, which is mediated by a glycosyl carrier lipid.

The mechanism of chain elongation in O-antigen synthesis is therefore analogous to that in fatty acid and polypeptide synthesis, and requires that the reducing terminus of the nascent polymer chain remain in an activated form throughout the entire sequence of glycosyl transfer reactions (Fig.71). This mechanism of polymerization also makes excellent sense in terms of the molecular architecture of the cell membrane; the lipid-linked reducing end of the growing polysaccharide chain can be held in close apposition to the polymerizing enzyme (reaction 5, Fig.68) and the incoming monomer unit within the organized lipid-rich membrane structure, while the non-reducing end of the polymer is free to extend into the more hydrophilic environment external to the membrane surface and at a distance from the active site of the polymerase.

JOINING THE PARTS

The last reaction in the synthesis of LPS is the linking of the completed O-side chain to the core. This unusual reaction, involving the joining of two macromolecules, was found to occur in a cell free system. O-Side chains linked to PP-GCL were generated *in vitro* in a cell envelope preparation from a core-defective mutant, and this particulate preparation was incubated with an exogenous core LPS. Efficient transfer of the O-side chain to the core was catalyzed by an enzyme or enzymes present in the cell envelope fraction. Unfortunately, since a satisfactory method for the extraction of O-side chain polysaccharide linked to PP-GCL is not yet available, further resolution of this system has not been achieved.

LPS can also be acetylated, as in the case of the LPS from *S. anatum* (see Fig.60). Cell envelope fractions of this organism were found to transfer acetyl groups from acetyl-CoA to galactose residues of endogenous LPS, or of exogenously added oligosaccharides, probably at C-6. The stage of synthesis of LPS at which the acetylation occurs in intact cells is not yet known.

Further light was shed on the biosynthesis of lipopolysaccharide by the isolation of "semirough" (SR) mutants, which possess O-specific chains in a reduced amount, as evidenced by their reduced capacity for binding the corresponding antibody. In one of these groups, $rouC$ mutants, the R antigen is entirely covered, presumably by a layer of single, non-repeating O-specific chains. This pattern suggests that the enzyme for transfering the O-specific chains to the R core (reaction 6, Fig.68) differs from the enzyme catalyzing the polymerization of the repeating unit (reaction 5, Fig.68), which must recognize a different acceptor group; the mutant presumably lacks the second of these enzymes.

Another kind of semirough mutant, $rouD$, has a similar small amount of O-antigen, but does not expose the underlying R-antigen as well. This mutant is believed to have a few specific side chains of normal length; presumably some early transfer enzyme is defective but still somewhat active ("leaky block").

ASSEMBLY OF THE OUTER MEMBRANE

Attention has recently been focused on the topological aspects of LPS biosynthesis and on the mechanism of assembly of the outer membrane, in which LPS is an essential constituent (A4). M. J. Osborn and her coworkers at the University of Connecticut have developed for this purpose a new technique for the isolation of inner (cytoplasmic) and outer membranes of *Salmonella*. The enzymes of O-antigen biosynthesis were shown to be entirely in the cytoplasmic membrane. The same location appears probable for the enzymes that synthesize the lipopolysaccharide core.

To obtain insight into the pathway of LPS assembly, a mutant of *S. typhimurium* was used which lacks phosphomannose isomerase, and is thus unable to form mannose from glucose or related carbon sources. In such a mutant, exogenous mannose is incorporated specifically into the O-specific chains. Pulse-chase experiments carried out with this mutant using radioactive mannose, showed that the label initially appeared in the LPS of the inner membrane but within two minutes was transferred to the outer membrane. This was also found to be true for both core sugars and for the O-antigen chains. Pre-existing core chains present in the outer membrane could not act as acceptor for newly synthesized O-antigen side chains.

The conclusions drawn from these experiments are that LPS is synthesized on the inner or cytoplasmic membrane, anchored to this membrane by the hydrophobic lipid A, and then translocated to the outer membrane. The minimal size of the translocated unit is not known, and the mechanism of translocation is far from being clear.

Electron microscope observations of *E. coli* have revealed the presence of a limited number of zones of adhesion or "bridges" between the inner and outer membrane. It has been assumed that translocation of LPS takes place over such bridges. This assumption is supported by the finding (A5) obtained with a UDP-Gal-4-epimerase deficient mutant of *S. typhimurium*, that newly synthesized LPS enters the outer membrane via discrete export sites. Such mutants, you may recall, will synthesize a complete LPS when galactose is added to their growth medium. After 30 seconds of incubation with galactose, about 220 patches of newly made complete LPS were observed on the surface of each cell, while after 2-3 minutes of growth, LPS was randomized. This was established by examination of the cells in the electron microscope after they had been treated with ferritin-conjugated antibodies directed against complete LPS.

PHAGE CONVERSION OF O-ANTIGEN

One of the most critical factors in determining whether a bacteriophage will infect a cell is the ability of the phage to become adsorbed onto the bacterial cell(6,7). Adsorption requires the presence of specific receptor sites on the cell surface with which one or more components of the phage can interact. In the case of the enteric bacteria, the receptors are present in the lipopolysaccharides. Thus, phage ε^{15} that will specifically infect *Salmonella anatum* will bind with the purified LPS from this organism, but not with the LPS extracted from other bacteria that do not adsorb ε^{15}.

Following its adsorption to the cell surface, the virus will generally kill its host or modify it by introducing new biosynthetic pathways into the metabolic sequences of the infected cell. The genetic material of virulent viruses may take command of the biosynthetic machinery of the host, cause the rapid synthesis of viral nucleic acid and protein and bring about total destruction of the host cell. Many bacteriophages are, however, temperate; instead of killing the cell, the viral DNA (or RNA) will be integrated into the bacterial chromosome and will be carried on in the dormant or lysogenic state. Although such bacterial phages do not alter the growth rate or overall economy of the cell while in the lysogenic state, they often cause subtle changes in the biosynthetic activities of the cell.

In 1956, H. Uetake of Kyoto University made the remarkable observation that cells that were lysogenic for

ϵ^{15}, would not adsorb new ϵ^{15} particles. Lysogenic infection by the ϵ^{15} phage apparently modified the bacterial surface and made subsequent adsorption impossible. The chemical basis for this change and the underlying genetic mechanism were elucidated in the early 1960's by the brilliant studies of Robbins and his coworkers (6, 7). At the same time they also clarified the mechanisms of the chemical, enzymic and genetic changes induced in S. *anatum* infected by phage ϵ^{34}, either directly or after lysogenic infection with phage ϵ^{15}.

In the O-specific side chains of S. *anatum*, the Man-L-Rha-Gal repeating units are linked to each other by α-(1→6) linkages between galactose and mannose. In addition, the galactose residues carry acetyl substituents on their C-6. Infection with phage ϵ^{15} leads to the production of polysaccharide chains which lack O-acetyl substituents, and the Gal-Man linkages are of the β rather than the α configuration. Such a polysaccharide can no longer serve as receptor for phage ϵ^{15} but is now a receptor for phage ϵ^{34}; infection with ϵ^{34} results in the glucosylation of the β-linked galactose residues. If S. *anatum* is infected with ϵ^{34} without prior infection with ϵ^{15}, no change in the polysaccharide is observed. However, if such cells are now infected by ϵ^{15}, the polysaccharide produced will be of the same structure as that obtained by the previously described double infection sequence. The alterations in O-antigen structure associated with conversion by ε^{15} and ε^{34} are summarized in Fig. 72.

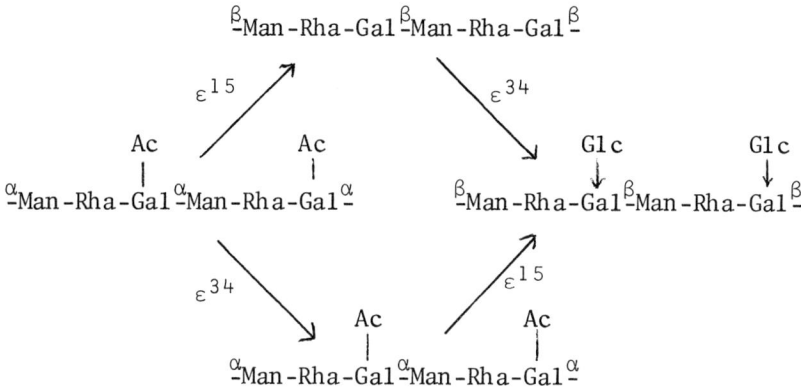

Figure 72 Conversion of the O-antigen of *S. anatum* by phages ε^{15} and ε^{34}.

Further analysis of the system has revealed that four phage specific functions or genes are responsible for these alterations: (a) repression of the host transacetylase which catalyzes acetylation of galactosyl residues by acetyl-CoA; (b) inhibition of the host α-polymerase, responsible for polymerization of the Man-L-Rha-Gal repeating units (reaction 5, Fig.68); (c) induction of a phage-specified β-polymerase; and (d) addition of glucose.

It is interesting to note that two different mechanisms of inactivation are employed here. The transacetylase is repressed, i.e. its synthesis is blocked; this is shown by the fact that after infection there is no increase in transacetylase activity, but the enzyme that was previously synthesized continues to show activity. The α-polymerase, on the other hand, continues to be synthesized by the cell, but is inactivated by a newly formed specific protein inhibitor. As a result, activity of the α-polymerase rapidly decreases, or ceases altogether, after infection. As mentioned, infection of *S. anatum* lysogenic for ε^{15} with ε^{34} results in an additional modification of the O-specific chains, namely glucosylation of their β-linked galactosyl residues. Indeed, the cell envelope fraction of such infected cells was shown to contain a new glucosyltransferase which catalyzed the addition of these glucosyl branch units from UDP-Glc to endogenous O-antigen acceptors. The mechanism of glucosylation was elucidated by A. Wright at Tufts University, Boston, who found that a lipid-linked derivative of glucose, Glc-1-P-GCL participates as an intermediate in the reaction (A2). The evidence strongly supports the following reaction sequence:

$$\text{UDP-Glc} + \text{P-GCL} \xrightarrow{\text{Mg}^{++}} \text{Glc-1-P-GCL}$$

$$\text{Glc-1-P-GCL} + (\text{Man-}\underline{\text{L}}\text{-Rha-Gal-}\beta\text{-})_n \rightarrow (\text{Man-}\underline{\text{L}}\text{-Rha-}\overset{\overset{\text{Glc}}{|}}{\text{Gal}}\text{-}\beta\text{-})_n$$

The studies of Robbins and his coworkers have clearly established that the O-antigenic structures are under genetic control, and that the presence or absence of specific enzymes, rather than a template mechanism, accounts for structural specificity in polysaccharides. In addition to their great impact on our knowledge of lipopolysaccharide structure and activity, they are of interest because there are significant similarities between the way phage ε^{15} or ε^{34} alters the surface of bacterial cells, and the action of tumor viruses on animal cells. Thus, the chromosome of tumor viruses may persist in the infected animal cell in a dormant stage, integrated into the host cell's chromosome. This dormant stage may be analogous to the lysogenic state of such phages as ε^{15} and ε^{34}. Moreover, transformed animal cells are known to have altered surface characteristics. Among the most prominent of these are modified carbohydrate receptors, as demonstrated by studies with lectins. The analogy between transformation of bacterial and animal cells should, however, be treated with caution, since many differences between the two processes are known. More work is required to establish further similarities and differences between the mechanisms by which a phage alters the surface of its bacterial host and the more complex mechanisms by which a tumor virus affects animal cells.

METABOLIC CONTROL OF LPS BIOSYNTHESIS

An important problem we have not considered is the mechanism of control of LPS synthesis at the enzymic level. Such a control may occur either through the regulation of the production of precursor sugar nucleotides, or through effects on the activity of the transferase enzymes. For a discussion of this subject, I refer you to the excellent review of Nikaido (1).

REFERENCES

Reviews

1. Biosynthesis of cell wall lipopolysaccharide in Gram-negative enteric bacteria,
   H. Nikaido, Advan. Enzymol. 31, 77-124 (1968).

2. Biosynthesis of saccharides from glycopyranosyl esters of nucleoside pyrophosphates ("sugar nucleotides"),
   H. Nikaido and W. Z. Hassid, Adv. Carb. Chem. Biochem. 26, 351-482 (1971).
   Comprehensive and detailed; deals with many aspects of the subject - from the synthesis of sugar nucleotides, through that of simple oligosaccharides, to highly complex carbohydrate polymers.

3. The role of membranes in the synthesis of macromolecules, M. J. Osborn, *in* Structure and Function of Biological Membranes (Ed. L. I. Rothfield), Academic Press, 1971, pp. 343-400.
   Deals with biosynthesis of lipopolysaccharides and other bacterial polysaccharides, of peptidoglycan and glycoproteins. Very interesting and highly recommended.

4. Biochemistry of bacterial cell envelopes,
V. Braun and K. Hantke, Ann. Rev. Biochem. 43, 89-121 (1974).

5. Metabolism and function of polyisoprenol sugar intermediates in membrane-associated reactions,
W. J. Lennarz and M. G. Scher, Biochim. Biophys. Acta 265, 417-441 (1972).

6. Direction of chain growth in polysaccharide synthesis,
P. W. Robbins, D. Bray, M. Dankert and A. Wright, Science 158, 1536-1542 (1967).

7. The receptor site for a bacterial virus,
R. Losick and P. W. Robbins, Scient. Amer. 221 (5), 120-126 (1969).
Excellent general presentation of the subject.

8. The relation of bacteriophage attachment to lipopolysaccharide structure,
A. M. C. Rapin and M. M. Kalckar, *in* Microbial Toxins (Eds. G. Weinbaum, S. Kadis and S. J. Ajl), Bacterial Endotoxins Vol.4, Academic Press, 1971, pp.267-307.

Specific articles

A1. Chemical synthesis of pyrophosphodiesters of carbohydrates and isoprenoid alcohols. Lipid intermediates of bacterial cell wall and antigenic polysaccharide biosynthesis,
C. D. Warren and R. W. Jeanloz, Biochemistry 11, 2565-2572 (1972).

A2. Isolation and characterization of nonconverting mutants of bacteriophage ϵ^{34},
A. Wright and N. Barzilai, J. Bacteriol. 105, 937-939 (1971).

A3. Glucosylation of lipopolysaccharide in *Salmonella*: biosynthesis of O antigen factor 12_2. II. Structure of the lipid intermediate,
K. Nikaido and H. Nikaido, J. Biol. Chem. 246, 3912-3919 (1971).
See also papers on pp.3902-3911, 3920-3927.

A4. Structure and biogenesis of the cell envelope of Gram-negative bacteria,
M. J. Osborn, P. D. Rick, V. Lehmann, E. Rupprecht and M. Singh, Ann. N.Y. Acad. Sci. 235, 52-65 (1974).

A5. Outer membrane of *Salmonella*. Sites of export of newly synthesised lipopolysaccharide on the bacterial surface, P. F. Mühlradt, J. Menzel, J. R. Golecki and V. Speth, Eur. J. Biochem. 35, 471-481 (1973).

A6. Bacteriophage attachment sites, serological specificity and chemical composition of the lipopolysaccharides of semirough and rough mutants of *Salmonella typhimurium*, A. A. Lindberg and C. G. Hellerqvist, J. Bacteriol. 105, 57-64 (1971).

PEPTIDOGLYCAN - I: STRUCTURE

Among the different constituents of the bacterial cell wall, the most important for the survival and integrity of the cell is the peptidoglycan (glycopeptide, mucopeptide or murein) (1-3; see also first part of lecture 17). It is found in all procaryotic organisms with the exception of those living under abnormal conditions, such as the halophilic bacteria which inhabit the Dead Sea, and the *Mycoplasma* which are normally intracellular parasites. Peptidoglycan is the polymer that imparts rigidity to the cell wall, and thus to the cell as a whole. Interference with its synthesis or its removal results in loss of this rigidity and leads to the formation of a fragile protoplast or spheroplast which under normal hypotonic conditions, swells and bursts.

Although considerable variations in the composition of

the cell wall occur when bacteria are grown under different conditions, the basic organization of the peptidoglycan, and most of its essential structural features, are not changed.

Our knowledge of peptidoglycan structure is derived largely from studies of the cell wall of a limited number of organisms: the Gram-positive *Micrococcus luteus* in which peptidoglycan comprises 60-70% of the dry weight of the cell wall, and *Staphylococcus aureus* (40-60% of the dry weight of the wall), and the Gram-negative *Escherichia coli* (5-10%). The first of these has been under investigation in our laboratory for over a decade.

Early studies on the composition of micrococcal and staphylococcal cell walls, carried out mainly by M. R. J. Salton in the 1950's, showed that, like the walls of many other Gram-positive bacteria, they contained only a few amino acids in significant concentration: usually alanine, glutamic, glycine and lysine. That these were not present as constituents of proteins was suggested by both their restricted range and by the discovery, made almost 20 years ago, that some of the amino acids were of the "unnatural" $\underline{D}$-configuration. The other major components found in acid hydrolysates of these cell walls were two amino sugars, glucosamine and muramic acid. The latter is the 3-O-$\underline{D}$-lactyl ether of glucosamine (see p. 16). A major advance in the elucidation of the nature of cell wall peptidoglycan came when J. T. Park found, in 1949 at the University of Wisconsin, that penicillin-inhibited *S. aureus* cells accumulated a previously unknown complex uridine derivative. The great significance of this finding, made almost simultaneously with the isolation by Leloir's group of the first sugar nucleotide, UDP-Glc, was not recognized at the time. The structure of these nucleotides, shown to contain N-acetylmuramic acid and the same amino acids (with the exception of glycine) as are present in the Staphylococcal cell wall, was elucidated several years later by Park and by J. L. Strominger. Since there was already evidence that penicillin kills bacteria by interfering with cell wall synthesis in growing cultures, Park and Strominger proposed, in a classical paper published in 1957 (A1), that the N-acetylmuramyl-peptide nucleotides might be precursors of the cell wall peptidoglycan, the biosynthesis of which was inhibited by penicillin. The peptidoglycan could thus be envisaged as consisting of a polysaccharide made of glucosamine and muramic acid residues, in their N-acetylated form, with peptide side chains attached to the muramic acid by an amide linkage.

Direct information on the structure of peptidoglycan

has been gained largely by the study of low molecular weight degradation products isolated from its enzymic digests (2). Most useful for this purpose was lysozyme from hen egg-white; but other glycosidases, as well as special peptidases, also played an important role in the elucidation of the structure of peptidoglycan.

STRUCTURE OF THE GLYCAN MOIETY

The first evidence that the glycan is a linear chain in which the component acetamido sugars alternate regularly, was obtained by M. R. J. Salton and J. M. Ghuysen in 1959. In their studies, they used cell walls of *M. luteus*, the substrate traditionally employed for the investigation of hen egg-white lysozyme. A typical preparation of purified cell walls of this organism contains the following constituents (molar proportions relative to glutamic acid): $\underline{D}$-glutamic acid 1, $\underline{L}$-lysine 1, glycine 1, $\underline{L}$-alanine 1, $\underline{D}$-alanine 1.2, N-acetylglucosamine 1, N-acetylmuramic acid 1, glucose 3, N-acetylaminomannuronic acid 1. (The last two sugars are part of a cell wall polysaccharide distinct from peptidoglycan; see p.328).

From lysozyme digests of *M. luteus*, Salton and Ghuysen isolated in small amounts a disaccharide, N-acetylglucosaminyl-

N-acetylmuramic acid (GlcNAc-MurNAc) and the corresponding tetrasaccharide, GlcNAc-MurNAc-GlcNAc-MurNAc. On the basis of chemical and enzymic experiments, the structure GlcNAc-β-(1→6)-MurNAc was proposed for this disaccharide. The correct structure of the disaccharide was established in 1963 by R. W. Jeanloz, H. M. Flowers and myself at the Massachusetts General Hospital in Boston (A2), after we had developed methods for its isolation on a large scale (e.g. 0.1 g from about 1.5 g of purified walls) and Flowers and Jeanloz had synthesized the disaccharide GlcNAc-β-(1→6)-MurNAc for comparison. Although the natural and synthetic products were very similar, if not identical, in most of their properties, they differed markedly in the molar color yield they gave in the Morgan-Elson reaction for N-acetylhexosamines. We concluded, therefore, that the natural disaccharide is β-(1→4) linked (Fig.73,I). Final proof for this structure was recently obtained by complete chemical synthesis of the fully acetylated derivative of the disaccharide GlcNAc-β-(1→4)-MurNAc, and the demonstration of its identity with the same derivative of the natural product (A3).

The tetrasaccharide, also isolated in relatively large amounts, has the structure GlcNAc-β-(1→4)-MurNAc-β-(1→4)-GlcNAc-β-(1→4)-MurNAc. The latter compound was one of the

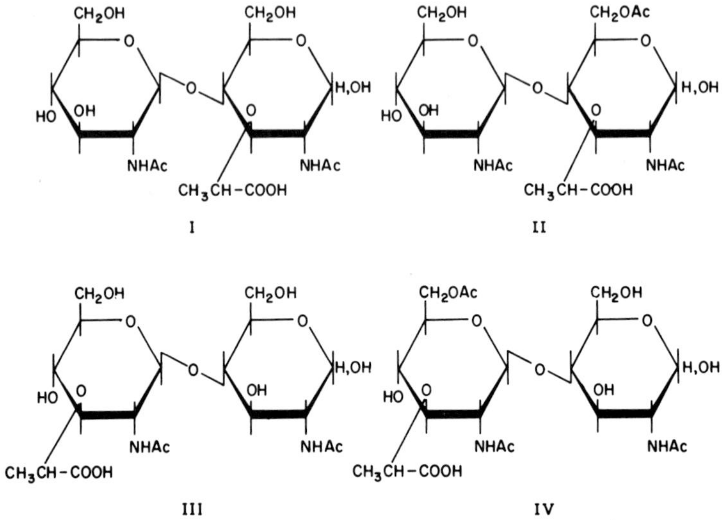

Figure 73 Structure of disaccharides isolated from the peptidoglycan of *M. luteus* (I, III) and of *S. aureus* Copenhagen (I-IV).

first low molecular weight compounds shown to be a substrate for lysozyme (Fig. 74). Subsequently, it played an important role, through the extensive work carried out in our laboratory (4) in helping to understand the mechanism of action of this enzyme (5).

In addition to providing much information on the structure of the glycan chain of peptidoglycan, formation of these oligosaccharides by lysozyme digestion of *M. luteus* shows that in its peptidoglycan a significant proportion (50%) of the *N*-acetylmuramic acid residues are devoid of

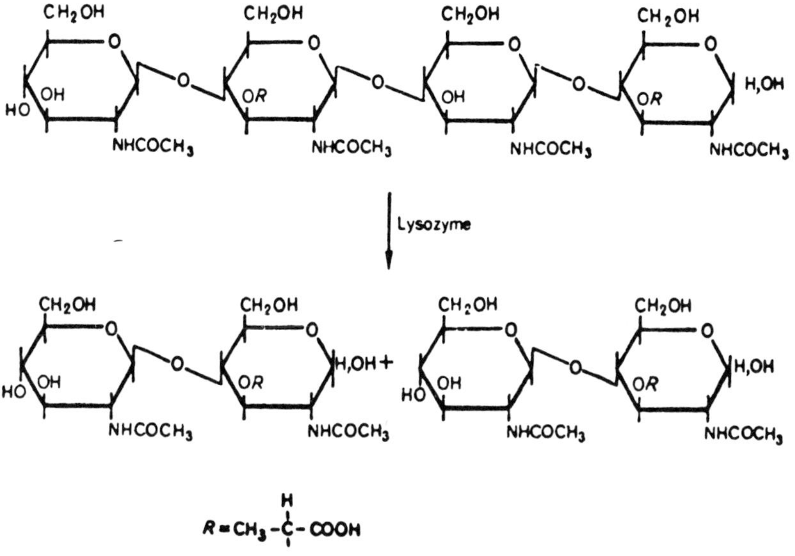

Figure 74 Hydrolysis of the cell wall tetrasaccharide by hen egg-white lysozyme to give the disaccharide GlcNAc-β-(1→4)-MurNAc.

peptide substituents. This is not a typical case; for the isolation of similar oligosaccharides from other organisms, it is necessary to digest their peptidoglycan both with lysozyme and a specific amidase that breaks the amide bond linking the peptide side chains with the lactyl group of N-acetylmuramic acid. Thus, lysis of cell walls of *S. aureus* Copenhagen with a lysozyme-like lytic enzyme (endo-N-acetyl-muramidase) from *Streptomyces albus* or from the fungus *Chalaropsis*, resulted in the hydrolysis of 95% of the N-acetyl-muramyl linkages, giving a high molecular weight glycopeptide consisting of crosslinked peptides substituted by disaccharide residues derived from the glycan chain. Removal of the peptide was effected with a *Streptomyces* N-acetylmuramyl-L-alanine amidase, which cleaved the amide linkage mentioned earlier, liberating N-terminal L-alanyl residues and two different disaccharides. One of these was identical with the disaccharide GlcNAc-β-(1→4)-MurNAc isolated earlier from *M. luteus* (Fig.73,I), while the other was a 6-O-acetyl derivative of this disaccharide, the substituent being on the N-acetylmuramic acid moiety (Fig.73,II).

The fragments released from peptidoglycan by hen egg-white lysozyme and related enzymes all contain N-acetylmuramic acid at their reducing end. Such enzymes, sometimes classified as N-acetylmuramidases, cannot cleave glycosidic linkages between N-acetylglucosamine and N-acetylmuramic acid. The latter linkage can be cleaved enzymically by special endo-N-acetylglucosaminidases, of which lysostaphin, isolated from a strain of *S. aureus*, is the best known representative. Lysostaphin preparations are, however, impure, and in addition to endo-N-acetylglucosaminidase, they contain endopeptidases, including an N-acetylmuramyl-L-alanine amidase. Prolonged incubation of *S. aureus* cell walls with lysostaphin led directly to the formation of two disaccharides in high yield. These were shown to possess the structure MurNAc-β-(1→4)-GlcNAc and 6-O-AcMurNAc-β-(1→4)-GlcNAc (Fig.73,III and IV). They are the corresponding isomers of the disaccharides isolated from lysozyme digests of cell walls. The disaccharide MurNAc-β-(1→4)-GlcNAc has also been found in lysostaphin digests of *M. luteus* cell walls.

Isolation of intact glycan from the walls of *S. aureus* Copenhagen has been achieved through the use of an extracellular enzyme from a *Myxobacter* species. This enzyme is a peptidase which hydrolyses D-alanylglycine, glycylglycine and *N*-acetylmuramyl-L-alanine bonds. The enzyme is devoid of glycosidase (glycanase) activity so that the undegraded glycan is also produced. Part of the glycan remains in association with the teichoic acid present in cell walls of *S. aureus* and, after removal of this complex by precipitation with trichloroacetic acid, glycan preparations essentially free from peptide were obtained. Such preparations had a low specific rotation, consistent with the presence of β-glycosidic linkages, and were resistant to oxidation with periodate. The isolated glycan was rapidly and completely hydrolyzed to disaccharides by the lysostaphin endo-*N*-acetylglucosaminidase. Treatment of the intact chains with tritiated sodium borohydride followed by hydrolysis of the reduced glycan afforded tritium labeled glucosaminitol but no radioactive muramicitol (the corresponding alcohol of muramic acid), showing that the glycan chains terminate at the reducing ends exclusively in N-acetylglucosamine. Chain length determination, based on either measurements of tritium incorporated on reduction with labeled borohydride or of formaldehyde formed on oxidation of the reduced glycan with periodate, indicate that the average chain length is about 12 to 16 disaccharide units, and the polymer is highly disperse, containing fractions of as few as 2 and as many as 50 disaccharide units. The factors which determine the glycan chain length are not known; it is possible that part of the terminal groups are formed by the action of endogenous, cell wall bound glycosidases. Indeed, *S. aureus* Copenhagen has been shown to contain an autolytic endo-N-acetylglucosaminidase which is presumably active during growth, cleaving the peptidoglycan either in the intact organism or in the isolated cell wall, thus accounting for the terminal reducing residues.

Approximately 60% of the N-acetylmuramic acid residues along the glycan chain of *S. aureus* Copenhagen are substituted at their C-6 by an O-acetyl group, but nothing is known of the distribution of the N,O-diacetylmuramic acid residues along the glycan chain.

Increasing evidence is becoming available on the structure of the glycan moiety of peptidoglycan in other bacteria (6). Purified peptidoglycan from most bacteria is susceptible to the action of lysozyme and related enzymes, all of which are believed to be specific for β-(1→4) N-acetylhexosaminidic linkages [although in our laboratory we have found that lysozyme will cleave GlcNAc-β-(1→2) linkages (A4)]. Furthermore, whenever disaccharides were isolated from cell wall digests, they appeared to be identical with GlcNAc-β-(1→4)-MurNAc, or MurNAc-β-(1→4)-GlcNAc. These and other findings strongly suggest that in all bacteria the glycan chains are similar to those found in *M. luteus* and *S. aureus* in that they consist of alternate residues of N-acetylglucosamine and N-acetylmuramic linked together by β-(1→4) bonds.

A small number of exceptions to the above structure has been reported. These include (a) the presence of a small amount of the manno isomer of muramic acid, instead of the usual glucose derivative; (b) replacement of the N-acetyl group of muramic acid by the N-glycolyl (CH_2OH-CO-) group (e.g. in *Mycobacteria*); (c) absence of a substituent on the amino group of muramic acid (A5) or of glucosamine; and (d) substitution of the C-6. No galactomuramic acid could be identified in any of the cell walls analyzed. It thus appears that the linear arrangement of the β-(1→4) linked chitin-like structure confers to the glycan backbone a conformation which is probably essential for the rigidity and function of the polymer, so that any mutation which would alter this conformation would probably be lethal. Epimerization at C-2 does not modify the basic conformation of the polymer, which explains the occurrence of small amounts of mannomuramic acid in walls of *M. luteus*.

STRUCTURE OF THE PEPTIDE MOIETY

In the case of *M. luteus*, the di- and tetrasaccharides are the principal components of the low molecular weight fraction obtained from lysozyme digests of the walls: together, these two products account for about 12% of the weight of the wall, or about 45% of its amino sugar content. Another important constituent of the digest are low molecular weight glycopeptides, in which N-acetylglucosamine and N-acetylmuramic acid are linked to peptide moieties consisting of D-glutamic, L-lysine, DL-alanine and glycine in the approximate molar ratios of 1:1:2:1, as found in the intact cell wall.

In 1967, D. Mirelman and I developed a preparative method which enabled us to isolate glycopeptides from lysozyme digest of *M. luteus* (A5) in quantities sufficient for complete chemical characterization. The major glycopeptide, present in the digest at about 10% of its weight, gave, upon digestion with *Streptomyces* amidase, the disaccharide GlcNAc-β-(1→4)-MurNAc and a pentapeptide. Partial acid hydrolysis of the pentapeptide afforded two fragments, L-Ala-D-Glu-Gly and L-Lys-D-Ala. The structure of these fragments, as well as that of the intact pentapeptide, was established by end group analysis and chemical degradation. In particular, hydrazinolysis experiments have shown that in the tripeptide the glycine is linked to the α-carboxyl of glutamic acid whereas the γ-carboxyl is free, and that in the intact pentapeptide the latter group is linked to the α-amino group of L-lysine, the ε-amino group of which is free. The structure of the disaccharide-pentapeptide based on these studies is given in Fig. 75. Another glycopeptide isolated by Mirelman proved to be a dimer of the disaccharide-pentapeptide, linked through a peptide bond formed between the free carboxyl of the terminal D-alanine of one peptide chain and the ε-amino group of the other peptide

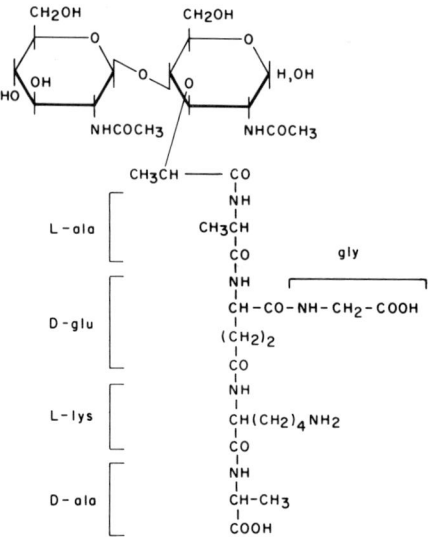

Figure 75 Structure of disaccharide-pentapeptide isolated from lysozyme digests of *M. luteus* cell walls (A5).

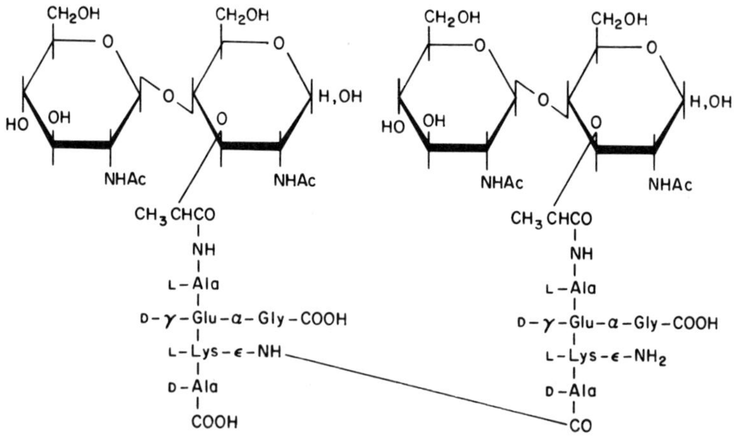

Figure 76 Structure of the disaccharide-pentapeptide dimer isolated from lysozyme digests of *M. luteus* cell walls (A5).

moiety (Fig.76).

The peptide sequence L-Ala-D-isoGlu-L-Lys-D-Ala is known to occur in nucleotide intermediates of bacterial cell wall synthesis; this structure has been established by chemical synthesis. The same peptide sequence has been shown to be present in cell walls of *S. aureus*.

In attempting to formulate the structure of intact *M. luteus* peptidoglycan from the fragments isolated after its digestion by lysozyme, two important facts must be borne in mind: (a) the chemical composition of the wall (p.396) seems to indicate the presence of one pentapeptide for each N-acetylmuramic acid residue; and (b) enzymic and chemical investigations (e.g. isolation of di- and tetrasaccharides) showed that at least 45% of the N-acetylmuramic acid residues are unsubstituted. To account for these apparently conflicting findings, it was postulated that in *M. luteus* peptidoglycan a second type of peptide bridge is present, in addition to the D-Ala-ε-L-Lys bridge found in the disaccharide-pentapeptide dimer described earlier. According to this assumption, the peptide subunits were also crosslinked through D-alanyl-L-alanyl linkages, involving the C-terminus of one subunit and the N-terminus of another. This assumption was proved correct by the isolation of the D-Ala-L-Ala

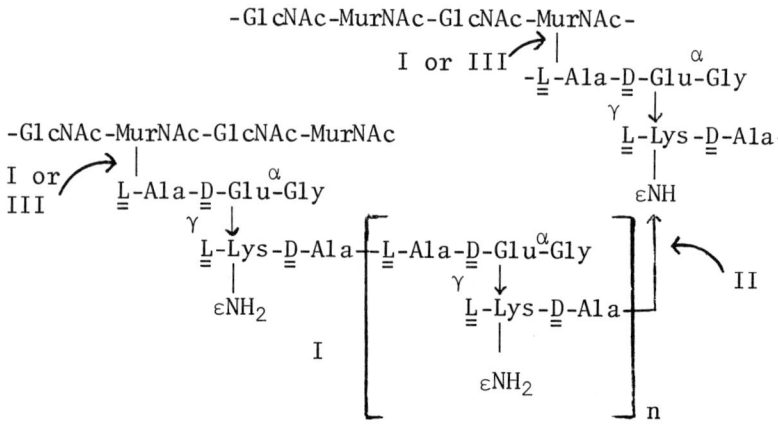

Figure 77 Peptide crossbridges in *M. luteus* peptidoglycan formed by pentapeptide subunits (n=2, 3 or 6). Arrows: peptide linkages hydrolyzed either by *Myxobacter* AL peptidase (I), by *Streptomyces* ML endopeptidase (II), or by N-acetylmuramyl-L-Ala amidase (III). In the course of wall degradation, the N-acetylmuramyl linkages are also hydrolyzed by accompanying glycosidases.

dipeptide from partial acid hydrolysates of *M. luteus*, and the pentapeptide oligomers, linked "head to tail" via D-Ala-L-Ala linkages, from walls digested with *Streptomyces* ML endopeptidase (which hydrolyzes the D-Ala-ε-L-Lys linkages, enzyme II in Fig.77) and an N-acetylmuramyl-L-Ala amidase (enzyme III, Fig.77).

The two types of peptide crosslinkages described interlink the glycan chains in *M. luteus* and lead to the formation of one giant highly insoluble bag-shaped macromolecule.

Extensive studies of the cell wall of *S. aureus* Copenhagen, using a variety of lytic enzymes, have established that the peptide moiety of its peptidoglycan consists of tetrapeptide side chains L-Ala-D-isoGlu-L-Lys-D-Ala, similar to those found in *M. luteus*. In *S. aureus*, however, the α-carboxyl of the glutamic acid is substituted by an amide group, and not by glycine as in *M. luteus*. The tetrapeptide side chains are crosslinked by pentaglycine bridges, extending from the D-alanine residue of one side chain, to the ε-NH$_2$ group of the L-lysine of another side chain (Figs. 78, 79).

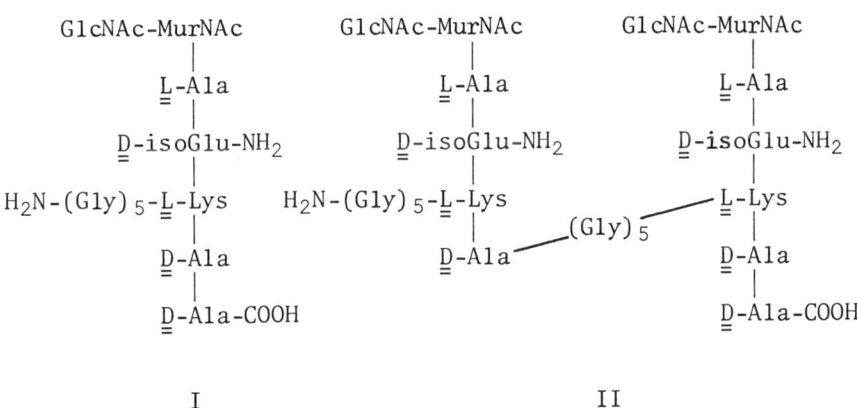

Figure 78 Glycopeptides obtained by digestion of *S. aureus* peptidoglycan with *N*-acetylmuramidase.
I: disaccharide-decapeptide; II: crosslinked glycopeptide.

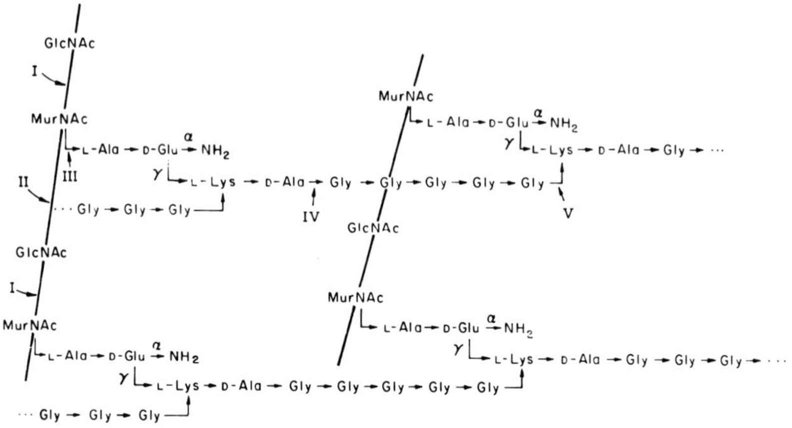
GlcNAc
I
MurNAc
L-Ala → D-Glu →α NH2
III
II
γ L-Lys → D-Ala → Gly → Gly → Gly → Gly → Gly
··· Gly → Gly → Gly
IV
V
GlcNAc
I
MurNAc
L-Ala → D-Glu →α NH2
γ L-Lys → D-Ala → Gly → ···
MurNAc
L-Ala → D-Glu →α NH2
γ L-Lys → D-Ala → Gly → Gly → Gly → ···
γ L-Lys → D-Ala → Gly → Gly → Gly → Gly → Gly
··· Gly → Gly → Gly

Figure 79 Structure of the peptidoglycan of *S. aureus* Copenhagen. Arrows: linkages cleaved by lysozyme (I), lysostaphin (II), *N*-acetylmuramyl-L-Ala amidase (III), *Myxobacter* AL peptidase (IV), SA endopeptidase (V). For further details see ref.2.

Confirmation of the structure of *S. aureus* peptidoglycan has been obtained using the *Myxobacter* peptidase which hydrolyzes mainly the pentaglycine bridge at three positions:

↓ ↓ ↓

-D-Ala-Gly-Gly-Gly-Gly-Gly-εNH-L-Lys-

giving tri- and tetraglycine and exposing *C*-terminal D-alanine.

Another important difference between the *M. luteus* peptidoglycan and that of *S. aureus* is that in the latter, all the carboxyl groups of the *N*-acetylmuramic acid residues are substituted by peptide side chains. This peptidoglycan possesses, therefore, a much tighter three-dimensional structure than that of M. luteus. It should be noted that the structure shown in Fig. 79 is an oversimplification in that the Staphylococcal cell wall is about 200 Å thick and so presumably contains several sheets of peptidoglycan held together by crosslinkages.

PEPTIDOGLYCAN CHEMOTYPES IN DIFFERENT BACTERIAL SPECIES

Whereas the glycan strands of peptidoglycan appear to be remarkably uniform in the great majority of bacterial species, this is not the case for the peptide moiety. The structure of the peptide side chains is known to vary, although to a limited extent, while the nature of the cross-bridges linking the peptide side chains is highly variable. Mainly as the result of the variation in the interpeptide bridges, about 60 different types of peptidoglycan are known today.

In most organisms studied, the peptide unit linked to the lactyl moiety of the N-acetylmuramic acid consists of four amino acids, with the general sequence L-alanyl (or sometimes L-seryl or glycyl)-D-glutamyl-L-R_3-D-alanine, where

R_3 is a side chain of a diamino acid (Fig.79). In this tetrapeptide, with an alternating D-L sequence, the glutamyl linkage is always γ, the other linkages being always α. The α-carboxyl of the D-glutamic acid may either remain unsubstituted, be amidated (as in *S. aureus*) or aminoacylated (e.g. by glycine in *M. luteus*). The most common amino acids found in the third position from the *N*-terminal residue (L-R_3) are L-lysine and LL or meso-diaminopimelic acid (DAP) (as in *E. coli*, for example). Other rare amino acids are, however, also found in this position (Fig.80).

The crosslinking between two tetrapeptides of adjacent glycan strands always involves the carboxyl of a *C*-terminal D-alanine residue of one tetrapeptide and either the amino group at the end of the side chain of the L-R_3 diamino acid (e.g. the ε-NH_2 of L-lysine) or the α-carboxyl group of D-glutamic acid. The crossbridges are of different types and may be arranged as homo- or heteropeptides containing up to six amino acid residues. They have been used by Ghuysen to classify the various peptidoglycans into four chemotypes.

In chemotype I the bridge consists of a direct linkage between the *C*-terminal D-alanine of one tetrapeptide, and the amino group on the R_3 side chain. We have encountered one example of such linkage in the disaccharide-pentapeptide dimer of *M. luteus*. Another example is the D-alanyl-meso-

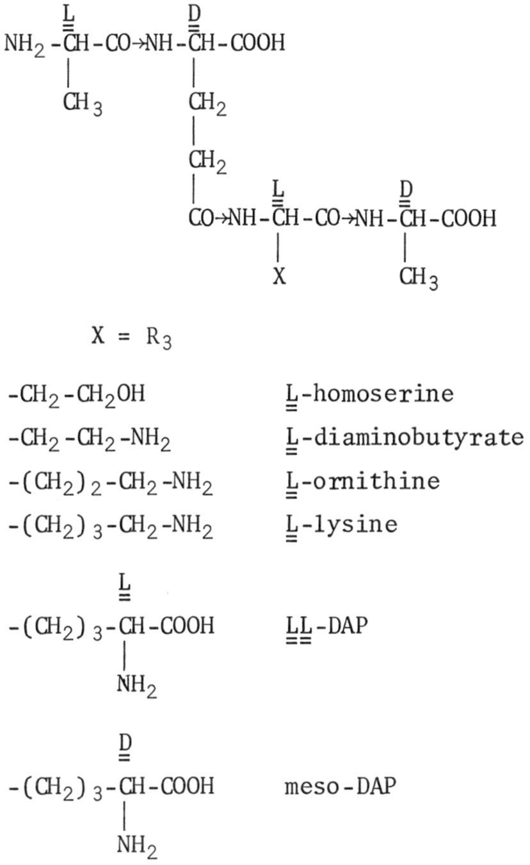

$$\text{NH}_2-\overset{\text{L}}{\underset{|}{\text{CH}}}-\text{CO}\rightarrow\text{NH}-\overset{\text{D}}{\underset{|}{\text{CH}}}-\text{COOH}$$

(Structure: L-Ala–D-Glu with γ-CH$_2$-CH$_2$-CO→NH-L-CH(X)-CO→NH-D-CH(CH$_3$)-COOH)

X = R$_3$

-CH$_2$-CH$_2$OH	L-homoserine
-CH$_2$-CH$_2$-NH$_2$	L-diaminobutyrate
-(CH$_2$)$_2$-CH$_2$-NH$_2$	L-ornithine
-(CH$_2$)$_3$-CH$_2$-NH$_2$	L-lysine
-(CH$_2$)$_3$-CH(L-NH$_2$)-COOH	LL-DAP
-(CH$_2$)$_3$-CH(D-NH$_2$)-COOH	meso-DAP

Figure 80 Structure of the tetrapeptide units. The main variations reside in the structure of the side chain of the L-R$_3$ residue. Sometimes the N-terminal amino acid is L-serine or glycine.

diaminopimelic acid linkage in the peptidoglycan of *E. coli* and *Bacillus*. In chemotype II, one additional amino acid or an intervening short peptide [(Gly)$_5$ in *S. aureus* or L-Ala-L-Ala in certain *Lactobacilli*] extends between the two tetrapeptides. In chemotype III, a variation of chemotype II, the bridge is made up from one or several peptides each having the same amino acid sequence as the peptide subunit. Such a crossbridge is found in *M. luteus* (Fig.77). Chemotype IV differs from the other chemotypes, since here one additional diamino acid (such as D-diaminobutyric acid, D-ornithine or D-lysine), or a short peptide containing a diamino acid, extends from the C-terminal alanine to the γ-carboxyl group of D-glutamic acid.

The tightness of the peptidoglycan net depends upon the length of the glycan strands, the frequency with which the glycan strands are substituted and the frequency with which the tetrapeptides are crosslinked. Depending upon the bacteria, the chain length of the glycan strands averages between 20 to 140 hexosamine residues, the percentage of peptide-substituted N-acetylmuramic acid varies from 50-100, and the average size of peptide moieties is between 2 and 10 crosslinked peptide subunits. Evidently, many terminal groups are present in both the glycan and the peptide moieties of the net. They reflect at least in part the dynamics of bacterial growth.

GENERAL STRUCTURE OF THE PEPTIDOGLYCAN

No direct study of the secondary structure of peptidoglycan has yet been made, but models have been described. In one of these (A6) glycan chains are aligned head to tail and are hydrogen-bonded together. A model in which the peptide chains adopt a helical configuration allows the formation of only a few hydrogen bonds. The repeating unit required for an α-helix is not present because of the involvement in peptide linkages of the γ-carboxyl group of glutamic acid and the ε-amino group of lysine. However, a pleated sheet (β-configuration) structure containing *trans* carboxyl groups provides a regular netlike structure in which 60% of the amino groups could be hydrogen-bonded. For more recent and interesting proposed configurations of peptidoglycan structure, shown by space-filling molecular models, see reference A7.

REFERENCES

Reviews

1. See lecture 17, references 1-4.

2. Use of bacteriolytic enzymes in determination of wall structure and their role in cell metabolism,
J. M. Ghuysen, Bacteriol. Rev. $\underline{32}$, 425-464 (1968).
An excellent review of peptidoglycan structure, written by one of the major contributors to our knowledge of this subject. Highly recommended.

3. The chemistry of Staphylococcal cell walls,
A. R. Archibold, in The *Staphylococci* (Ed. J. O. Cohen), Wiley-Interscience, New York, 1972, pp.75-109.
Although it focuses on a single organism, is very worth reading to anybody interested in cell walls in general, their structure (including teichoic acids), biosynthesis and mechanism of antibiotic action.

4. Mechanism of lysozyme action,
D. M. Chipman and N. Sharon, Science $\underline{165}$, 454-465 (1969).

5. Lysozyme,
E. F. Osserman, R. E. Canfield and S. Beychok, Editors, Academic Press, 1974, 645 pp.
Proceedings of a symposium commemorating the 50th anniversary of the discovery of lysozyme by Sir Alexander Fleming. Contains a wealth of information on the enzyme, its mechanism of action, as well as on the structure of bacterial cell wall peptidoglycan (see particularly articles on pp.9, 169, 185, 195 and 229). There is also a bibliography of 2600 papers on lysozyme published in the period 1922-1972.

6. Peptidoglycan types of bacterial cell walls and their taxonomic implications,
K. H. Schleifer and O. Kandler, Bacteriol. Revs. $\underline{36}$, 407-477 (1972).
Very comprehensive source of information.

Specific articles

A1. Mode of action of penicillin. Biochemical basis for the mechanism of action of penicillin and for its selective toxicity,
J. T. Park and J. L. Strominger, Science $\underline{125}$, 99-101 (1957).

A2. Isolation and the study of the chemical structure of a disaccharide from *Micrococcus lysodeikticus* cell walls, N. Sharon, T. Osawa, H. M. Flowers and R. W. Jeanloz, J. Biol. Chem. $\underline{241}$, 223-230 (1966).

A3. Synthesis of the repeating disaccharide unit of the glycan moiety of the bacterial cell wall peptidoglycan, C. Merser and P. Sinaÿ, Tetrahedron Lett. $\underline{13}$, 1029-1032 (1973).

A4. Formation and cleavage of 1→2 glycosidic bonds by hen's egg-white lysozyme,
J. J. Pollock and N. Sharon, Biochem. Biophys. Res. Commun. 34, 673-680 (1969).

A5. Isolation and study of the chemical structure of low molecular weight glycopeptides from *Micrococcus lysodeikticus* cell walls,
D. Mirelman and N. Sharon, J. Biol. Chem. 242, 3414-3427 (1967).

A6. Three-dimensional molecular models of bacterial cell wall mucopeptides (peptidoglycans),
M. V. Keleman and H. J. Rogers, Proc. Natl. Acad. Sci. USA 68, 992-996 (1971).

A7. Two proposed general configurations for bacterial cell wall peptidoglycans shown by space-filling molecular models,
E. H. Oldmixon, S. Glauser and M. L. Higgins, Biopolymers 13, 2037-2060 (1974).

PEPTIDOGLYCAN - II: BIOSYNTHESIS AND MODE OF PENICILLIN ACTION

In the previous lecture we have seen that peptidoglycan is basically a polymer of the repeating disaccharide-tetrapeptide, N-acetylglucosaminyl-N-acetylmuramyl-$\underline{L}$-Ala-$\underline{D}$-isoGlu-$\underline{L}$-Lys(or DAP)-$\underline{D}$-Ala. In the polymer, the GlcNAc-MurNAc disaccharide units are joined glycosidically to form linear glycan chains and the peptide units are at least partially crosslinked, either directly or through peptide bridges.

Since the shape of the bacterial cell is maintained largely by peptidoglycan, detailed understanding of the biosynthesis of peptidoglycan requires knowledge, not only of how the component amino sugars and amino acids become joined together, but also of the process by which the product is assembled in an orderly manner externally to the cytoplasmic membrane, and then modified during growth and cell division (1,2). Although very little is as yet known of the latter processes, many of the enzymic steps leading to the synthesis of the crosslinked peptidoglycan have now been established (3,4).

Present knowledge of the biosynthesis of this highly complex and unusual macromolecule is based mainly on extensive investigations carried out with *M. luteus*, *S. aureus* and *E. coli*. These investigations, begun over 25 years ago by J. T. Park, were continued by him, as well as by J. L. Strominger, F. C. Neuhaus in the U.S., by H. J. Rogers and H. R. Perkins in England, and by many other workers. The findings obtained clearly indicate that all known peptidoglycans are synthesized by common mechanisms, and adequately account for their common, as well as their different, structural characteristics.

The process of peptidoglycan biosynthesis is usually divided into three stages:

[1] Synthesis of the two sugar nucleotides that serve as the low molecular weight precursors of peptidoglycan

[2] Conversion of the water-soluble nucleotides into lipid-soluble bactoprenyl (or glycosyl carrier lipid) derivatives, introduction of appropriate alterations in the peptide units, and the subsequent polymerization of these activated deriv-

atives via a transglycosylation reaction to form linear peptidoglycan strands.

[3] Transfer of the polymerized subunit to acceptor sites on the cell wall and crosslinking of the linear peptidoglycan strands by a transpeptidation reaction to form a three-dimensional polysaccharide-peptide network.

Ample evidence on the reactions involved in the first stage is available, but little is known about the second and third steps.

SYNTHESIS OF PRECURSORS

The first stage consists of the synthesis of UDP-GlcNAc and of UDP-MurNAc-$\underline{L}$-Ala-$\underline{D}$-isoGlu-$\underline{L}$-Lys-$\underline{D}$-Ala-$\underline{D}$-Ala (UDP-MurNAc-pentapeptide, Fig.81), the sugar nucleotide precursors of peptidoglycan. In organisms that contain diaminopimelic acid (DAP) instead of $\underline{L}$-lysine, the pentapeptide moiety is $\underline{L}$-Ala-$\underline{D}$-isoGlu-DAP-$\underline{D}$-Ala-$\underline{D}$-Ala. UDP-MurNAc-pentapeptide, originally isolated by Park some 25 years ago, is the major sugar nucleotide that accumulates in large amounts in penicillin inhibited *S. aureus* cells. This is, however, an unusual case, since in most other organisms no such accumulation could be induced. The two other nucleotides, also first isolated in 1952 by Park from the same cells, were subsequently identified as UDP-N-

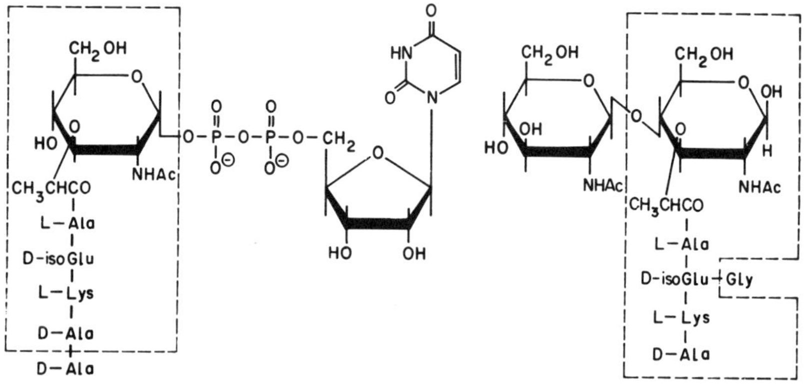

Figure 81 Structure of UDP-MurNAc-pentapeptide, a key
intermediate in the synthesis of peptidoglycan
in *S. aureus* and *M. luteus* (I). For comparison,
the structure of the disaccharide-pentapeptide
isolated from *M. luteus* cell wall is given (II).
Note in particular that the peptide side chain
in the completed peptidoglycan (II) ends in a
single D-alanine residue, whereas the precursor
contains the D-Ala-D-Ala sequence instead.

acetylmuramic acid and UDP-*N*-acetylmuramyl-L-alanine. It was subsequently found that the different uridine nucleotides also accumulate in cells of *S. aureus* under conditions of nutritional deficiency or on treatment with antibiotics other than penicillin (e.g. vancomycin or bacitracin). Thus, lysine deprivation leads to the accumulation of UDP-*N*-acetylmuramyl-L-alanine-D-glutamic acid, whereas treatment of the cells with the antibiotic D-cycloserine (a structural analog

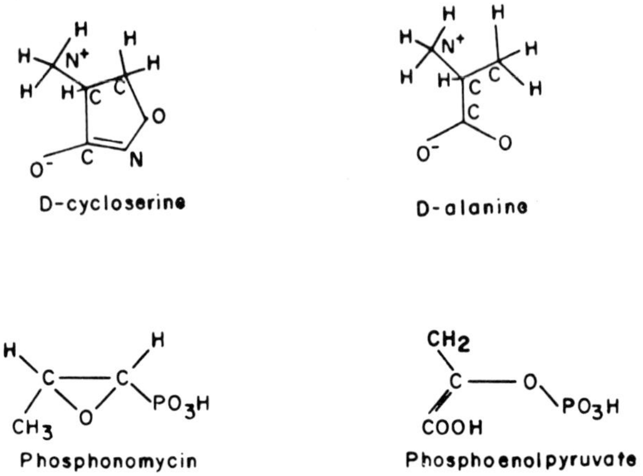

Figure 82 Structure of D-cycloserine and phosphonomycin, competitive inhibitors of D-alanine and phosphoenolpyruvate, respectively. D-Cycloserine and D-alanine are represented in their zwitterion forms.

of D-alanine, Fig.82), induces the accumulation of UDP-MurNAc-L-Ala-D-isoGlu-L-Lys. Gentian violet induced the accumulation of UDP-MurNAc, and phosphonomycin (a competitive inhibitor of phosphoenolpyruvate, Fig.82) the accumulation of UDP-GlcNAc. Numerous bacterial mutants deficient in enzymes involved in peptidoglycan synthesis have also been found to accumulate some of the above nucleotides. These compounds clearly represent a likely biosynthetic sequence leading from UDP-GlcNAc to the complex sugar nucleotide pentapeptide (Fig.83).

Using the various sugar nucleotides isolated from inhibited cells, Strominger and his coworkers established the reaction sequence leading to the formation of UDP-MurNAc-

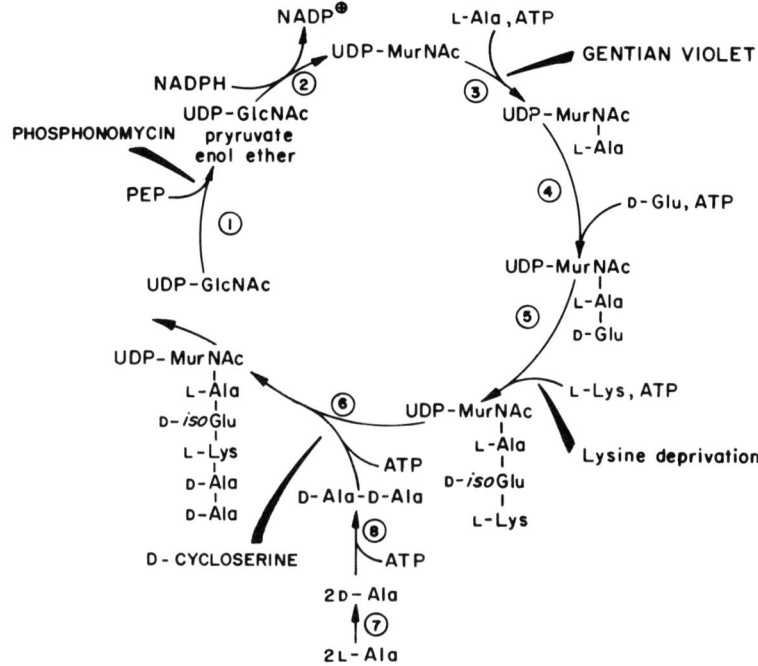

Figure 83 Biosynthesis of UDP-N-acetylmuramyl-pentapeptide in *S. aureus*. PEP=phosphoenolpyruvate.

pentapeptide in *S. aureus*. In brief, it is as follows: first, the condensation of N-acetylglucosamine 1-phosphate with UTP, catalyzed by a uridyltransferase (or pyrophosphorylase) leads to the formation of UDP-N-acetylglucosamine. A specific transferase then catalyzes a reaction of phosphoenolpyruvate (PEP) with UDP-GlcNAc, to give the 3-pyruvate enol ether of the latter (reaction 1, Fig. 83; Fig. 84). This transferase is irreversibly inhibited by phosphonomycin. Reduction of the pyruvyl group by an NADPH-linked reductase

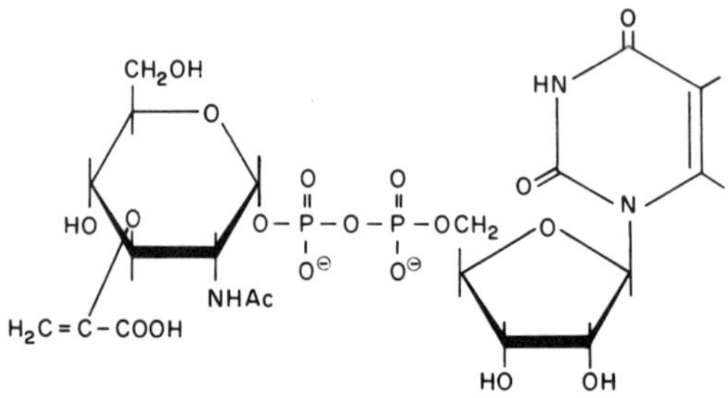

Figure 84 UDP-*N*-acetylglucosamine 3-pyruvate enol ether

leads to the formation of the 3-*O*-D-lactyl ether of *N*-acetyl-glucosamine (reaction 2); in this way UDP-*N*-acetylmuramic acid is obtained from UDP-GlcNAc. Conversion of UDP-MurNAc to its pentapeptide form occurs by the sequential addition of the requisite amino acids. Each step requires ATP, a divalent cation (Mg^{++} or Mn^{++}) and a specific enzyme which catalyzes the addition of the amino acids; the ATP is cleaved into ADP. The addition of L-alanine occurs first (reaction 3), followed by D-glutamic acid (reaction 4), later amidated in *S. aureus* to D-isoglutamine; L-lysine is then attached by its α-amino group to the γ-carboxyl group of the glutamic acid (reaction 5) and finally the dipeptide D-Ala-D-Ala is added as a preformed unit; the latter reaction is also coupled with the hydrolysis of ATP to ADP (reaction 6). This dipeptide is formed from L-alanine by two enzymic reactions:

conversion of L-alanine to D-alanine by a racemase (reaction 7), followed by the linking of the two alanine residues in an ATP-requiring reaction to form D-Ala-D-Ala (reaction 8). Both the alanine racemase and the D-alanyl-D-alanine synthetase are strongly and competitively inhibited by D-cycloserine, which explains why this antibiotic induces accumulation of UDP-MurNAc-L-Ala-D-isoGlu-L-Lys in *S. aureus* cells. Reactions 1-8, all of which occur in the soluble, cytoplasmic compartment of the bacterial cells, have been fairly well characterized. Whenever tested in other organisms, the sequence of reactions has been found to be essentially the same as in *S. aureus*.

Aminoacyl adenylates, which serve as intermediates in the biosynthesis of proteins, are apparently not involved in the formation of the peptide sequence of UDP-MurNAc-pentapeptide. As mentioned above, the addition of amino acids to form this sugar nucleotide-peptide is coupled with the cleavage of ATP into ADP and inorganic phosphate. Synthesis of the pentapeptide is thus quite distinct from protein synthesis, and the sequence of amino acids in the peptide is determined solely by the specificities of the synthetic enzyme. The separate enzymes involved in the addition of each amino acid (and of the dipeptide D-Ala-D-Ala) are highly specific for both the amino acid and the nucleotide substrate. For example, the purified D-glutamic acid-adding enzyme did not add glutamic acid to any nucleotide derivative other than UDP-MurNAc-L-Ala, nor did it add any other amino acid to the nucleotide.

ASSEMBLY OF THE DISACCHARIDE-PEPTIDE REPEATING UNIT

The second stage in the biosynthesis of peptidoglycan, first demonstrated in 1964, is catalyzed by membrane-bound enzymes. It is the polymerization of N-acetylglucosamine together with N-acetylmuramyl-pentapeptide to form the linear peptidoglycan strands. This reaction turned out to be far more complicated than was at first apparent. Initially it was believed that the glycan chain was formed by simple and direct transglycosylation involving the two precursors, UDP-GlcNAc and UDP-MurNAc-pentapeptide. However, when the reaction was examined in detail using a variety of radioactive precursors, the products turned out to be somewhat unexpected. Both UDP and UMP were formed in the polymerization reaction, the former compound was shown to be derived from UDP-GlcNAc, whereas the latter (UMP) was from UDP-MurNAc-pentapeptide. Formation of UMP was also accompanied by the release of inorganic phosphate. These findings were inconsistent with a direct transglycosylation reaction of the type involved, for example, in the biosynthesis of glycogen. The clue to this anomaly came from careful examination of the products found on paper chromatography of incubation mixtures containing the two substrates and the particulate enzyme. The polysaccharide formed does not migrate on paper and remains at the origin; there was, however, always a trace of radioactivity present at the solvent front. For some time this radioactive "contaminant" was ignored, but it eventually proved to be the key to the mechanism of the reaction. This contaminant was shown to be a lipid intermediate in the reaction sequence, eventually identified as a condensation product of N-acetylmuramyl-pentapeptide 1-phosphate and bactoprenyl phosphate (GCL-phosphate). This is, in fact, how bactoprenyl phosphate, now known to serve as an intermediate in the synthesis of other complex bacterial polysaccharides (e.g. the O-side chains of LPS; see lecture 19) was first encountered. Identification of bactoprenyl phosphate took several years, mainly because it represented only about 0.1% of the total phospholipid of the membrane.

In parallel with the studies that led to the identification of the lipid intermediate, the sequence of reactions

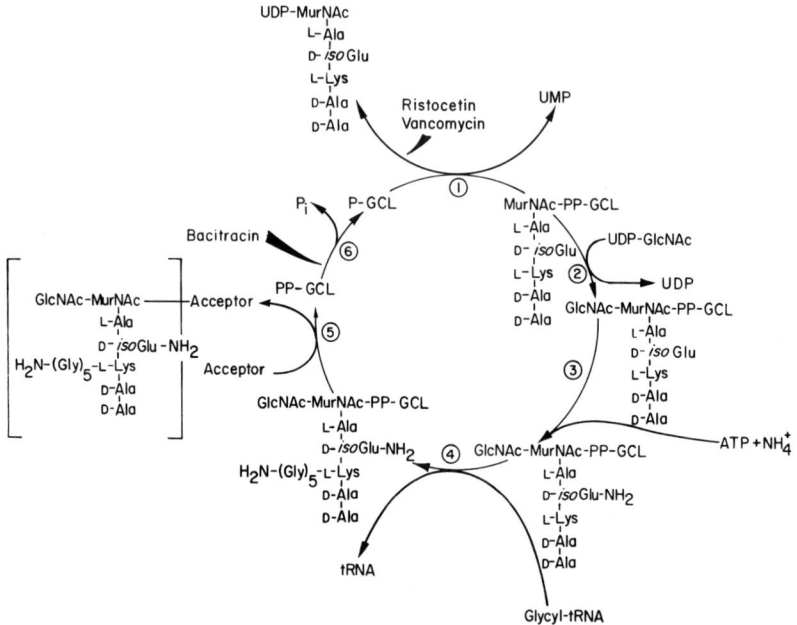

Figure 85 Pathway of biosynthesis of the peptidoglycan of
S. aureus. In E. coli the same reaction cycle
occurs, with the following exceptions: L-lysine
is replaced by meso-DAP; D-glutamic is not amidated, and no additional amino acid is incorporated into the pentapeptide side chain. Note
that at the end of the cycle the disaccharide-
peptide units are uncrosslinked.

leading to the formation of the disaccharide-pentapeptide repeating unit was worked out (Fig.85). The initial step, formation of the monosaccharide-pentapeptide-lipid intermediate mentioned earlier, occurs by reversible transfer of MurNAc-(pentapeptide) 1-phosphate from UDP-MurNAc-pentapeptide to bactoprenyl phosphate (P-GCL), with release of UMP (reaction 1, Fig.85).

Transfer of the entire MurNAc-phosphate portion of the sugar nucleotide to the acceptor lipid was demonstrated by use of ^{32}P-labelled UDP-MurNAc and the presence of a pyrophosphate linkage in the product was ultimately established by chemical degradation of the purified lipid-linked intermediate. Formation of MurNAc-(pentapeptide)-PP-GCL is, as mentioned, a readily reversible reaction. The high-energy character of the glycosyl-pyrophosphate bond of the sugar nucleotide is therefore fully conserved in the lipid-linked intermediate which acts as a glycosyl donor in the subsequent polymerization of the polysaccharide chain.

In the second step in the assembly of the peptidoglycan repeating unit, MurNAc-(pentapeptide)-PP-GCL acts as acceptor of *N*-acetylglucosamine from UDP-GlcNAc to form the disaccharide-pentapeptide intermediate (reaction 2, Fig.85). Addition of *N*-acetylglucosamine appears to occur by a conventional glycosyl transfer reaction, the nucleotide product of the reaction being UDP. This reaction is not readily reversible and the available evidence is consistent with the direct transfer of *N*-acetylglucosamine from the sugar nucleotide donor to the monosaccharide-lipid acceptor, without intervention of intermediates that can be isolated.

It has been proposed that lipid intermediates of peptidoglycan synthesis are formed in order that the hydrophilic precursors of peptidoglycan can be transported across the hydrophobic interior of the cell membrane. Such transport is necessary because peptidoglycan is located on the outer side of the cell membrane, whereas its UDP-linked precursors are formed in the cytoplasm. Furthermore, it seems likely that these intermediates can function as donors of glycosyl groups in a hydrophobic environment, either within or around the membrane, an area into which sugar nucleotides would be unable to penetrate.

MODIFICATION OF THE PEPTIDE SIDE CHAIN; ROLE OF tRNA

Membrane fractions of *S. aureus* and *M. luteus* are able to utilize the disaccharide-pentapeptide intermediate directly for synthesis of polymeric peptidoglycan. However, in these and other organisms in which the peptide subunits contain additional bridge amino acids and/or substituted D-glutamyl residues, the modifications are introduced prior to polymerization at the level of disaccharide-(pentapeptide)-PP-GCL. The α-carboxyl group of the D-glutamate residue is substituted in *S. aureus* by an amide group and in *M. luteus* by a single glycine residue.

Formation of the isoglutamine amide linkage by membrane fractions of *S. aureus* (reaction 3, Fig.85) required either NH_4^+ or glutamine and ATP; both the monosaccharide- and disaccharide-(pentapeptide)-PP-GCL intermediates were effective as substrates for amidation, but UDP-MurNAc-pentapeptide was totally inactive. Enzymic addition of the glycine residue by membrane fractions of *M. luteus* has also been shown to occur at the level of the lipid intermediates, according to the reaction:

$$\begin{array}{c}
\text{GlcNAc-MurNAc-PP-GCL} \\
| \\
\underline{L}\text{-Ala} \\
| \\
\underline{D}\text{-isoGlu-COOH} \\
\gamma | \\
\underline{L}\text{-Lys} \\
| \\
\underline{D}\text{-Ala} \\
| \\
\underline{D}\text{-Ala}
\end{array}
+ \left\{\begin{array}{c} \text{ATP} \\ + \\ \text{Gly} \end{array}\right\} \rightarrow
\begin{array}{c}
\text{GlcNAc-MurNAc-PP-GCL} \\
| \\
\underline{L}\text{-Ala} \\
| \\
\underline{D}\text{-isoGlu-CO-NHCH}_2\text{COOH} \\
\gamma | \\
\underline{L}\text{-Lys} \\
| \\
\underline{D}\text{-Ala} \\
| \\
\underline{D}\text{-Ala}
\end{array}
+ \left\{\begin{array}{c} \text{ADP} \\ + \\ P_i \end{array}\right.$$

The reaction was independent of tRNA and resulted in formation of ADP and inorganic phosphate (P_i), rather than AMP and pyrophosphate (PP_i). The mechanism of peptide bond formation is therefore quite different from that in tRNA-dependent synthesis of bridge peptide units (see below) and resembles rather the synthesis of glutathione and the pentapeptide unit itself.

The role of transfer RNA (tRNA) in the synthesis of the interpeptide bridge units of peptidoglycan was suggested by the initial observation of A. N. Chatterjee and J. T. Park made in 1964, that incorporation of glycine (but not of MurNAc-pentapeptide) into peptidoglycan in cell free preparations from *S. aureus* was abolished by ribonuclease. Subsequent studies by Strominger and his coworkers established the general mechanism whereby aminoacyl-tRNA's act as obligatory donors of amino acid residues in the assembly of bridge peptide units. It was shown unequivocally that glycyl-tRNA was the immediate precursor of the pentaglycine bridge unit in *S. aureus* and that the addition of glycine occurred exclusively at the level of the lipid-linked intermediates (reaction 4, Fig.85). Disaccharide-(pentapeptide)-PP-GCL and, to a lesser extent, the monosaccharide intermediate, were effective as acceptors of glycine, while preformed glycine-deficient polymeric peptidoglycan or UDP-MurNAc-pentapeptide was totally inactive.

Attachment of the oligoglycine chains to the ε-amino group of the lysine residue was established by dinitrophenylation and Edman degradation of the enzymic products. Elongation of the oligoglycine chain appears to occur by successive stepwise transfer of single glycine residues from glycyl-tRNA to the free amino terminus of the peptide. No evidence for intermediate formation of peptidyl-tRNA could be obtained.

Investigations in several other organisms having different bridge structures have confirmed the essential role of tRNA. Thus, the appropriate aminoacyl-tRNA's were found to serve as donors of L-serine and glycine in the synthesis of the mixed glycyl-seryl pentapeptide bridge in *Staphylococcus epidermidis*, and of L-threonine and L-alanine in the formation of the tetrapeptide bridge of *Micrococcus roseus*. In each case, addition of bridge units to the ε-NH_2 group of lysine occurred prior to polymerization, at the level of mono- and disaccharide-(pentapeptide)-PP-GCL intermediates.

Evidence that specificity in peptide bridge formation is at least partially determined by the tRNA molecule itself has been obtained by Strominger and his coworkers in studies of the role of L-alanyl-tRNA in the synthesis of the bridge in *Arthrobacter crystallopoietis*. L-Alanine linked to cysteine-specific tRNA (tRNA^{Cys}) was prepared by reduction of L-cysteinyl-tRNA with Raney nickel. The activity of the product, L-alanyl-tRNA^{Cys}, in peptide bridge synthesis was compared with that of L-alanyl-tRNA^{Ala}. The results showed that the L-alanyl-tRNA^{Cys} was completely inactive as donor of alanyl residues in peptidoglycan synthesis. The peptidoglycan synthesizing system thus appears to recognize some specific feature of the tRNA molecule. Whether this involves the amino recognition site or some other structural element of the tRNA is not yet known.

The important question of whether the tRNA species which participate in peptidoglycan synthesis are different from those involved in protein synthesis has been clarified in part by purification of the relevant tRNA's. Three glycyl-tRNA fractions from *S. aureus* were separated, all of which were approximately equally active in the peptidoglycan-synthesizing system. Two of the three were also active in polypeptide synthesis and were identified by triplet binding as corresponding to known glycine codons. The third fraction however failed to support polypeptide synthesis with natural or synthetic messengers and could not be identified with any of the known glycine anticodons by triplet binding.

Similar results were obtained on purification of the 4 L-seryl-tRNA's of *S. epidermidis*. Only three of the four were effective as donors of L-serine in synthesis of the bridge peptide of peptidoglycan. In contrast, purification of the L-threonyl-tRNA's of *M. roseus* yielded no evidence of a peptidoglycan-specific fraction. The occurrence in

Staphylococci of tRNA species apparently unique to peptidoglycan synthesis is of considerable interest, but the general occurrence of these and the specificity relationships *in vivo* remain to be established. It has been found that the degree of specificity of the bridge-synthesizing enzymes towards their aminoacyl-tRNA substrates may vary widely.

POLYMERIZATION OF DISACCHARIDE UNITS

Polymerization of the completed disaccharide-peptide repeating unit is believed to occur by successive transfer of disaccharide-peptide units from GCL to a suitable acceptor or primer (reaction 5, Fig.85). It should be emphasized that only the glycosyl moiety of the intermediate is transferred and that GCL is released in the form of a pyrophosphate derivative in which the terminal phosphate is derived from UDP-MurNAc-pentapeptide.

The final step which completes the reaction cycle depicted in Fig. 85 is dephosphorylation of the GCL-pyrophosphate by a specific bactoprenyl pyrophosphatase, to regenerate the GCL-phosphate. This reaction is specifically inhibited by low concentrations of bacitracin, a mixture of antibiotic peptides obtained from *Bacillus licheniformis*. Indeed, PP-GCL accumulates both in cell-free incubation mixtures and in intact cells, in the presence of this antibiotic. Interestingly, studies of the mode of action of bacitracin revealed that it acts not on the bactoprenyl pyrophosphatase, but by complexing with the substrate, GCL-pyrophosphate.

The nature of the acceptor or primer has not yet been investigated in detail, but it is possible that they are non-reducing termini of preexisting peptidoglycan chains. Thus, membrane fractions, obtained by dissolution of cell walls with N-acetylmuramidases or N-acetylhexosaminidases which degrade the endogenous glycan, lose the ability to incorporate lipid-linked intermediates into the polymer. In addition, incorporation into the polymer is inhibited by the antibiotics, vancomycin and ristocetin, which have been shown to bind tightly and specifically to the terminal end of peptide side chains. Stepwise addition of disaccharide-peptide units to end groups of glycan chains would fit well with current hypotheses of the mechanism of growth of cell wall, according to which controlled autolysis introduces nicks into peptidoglycan strands at defined growing points and provides new sites for chain elongation.

CROSSLINKING OF PEPTIDE UNITS

The last reaction in peptidoglycan synthesis is the

formation of crosslinks between peptide moieties. Studies of this reaction contributed greatly to the understanding of the mechanism of penicillin action.

It has been recognized for many years that the sugar nucleotide precursor of peptidoglycan, UDP-MurNAc-pentapeptide, contains two D-alanine residues at the carboxyl terminus of the peptide while the subunit of peptidoglycan usually contains only a single D-alanine residue (Fig.81). It was originally postulated by H. H. Martin in 1964 that the crosslinking reaction might involve transpeptidation between the D-alanyl-D-alanine terminus of one peptide unit and the free amino group of the basic amino acid or bridge unit of a second peptide, with release of free D-alanine (Fig.86). This hypothesis was particularly attractive in that it provided a mechanism for isoenergetic formation of the crosslinking peptide bond in the wall itself, external to the cytoplasmic membrane and independent of energy donors such as ATP. *In vivo* studies of E.M. Wise and J.T. Park (1965) and of D.J. Tipper and J.L. Strominger (1965,1968) in *S. aureus* provided support not only for the existence of such a transpeptidation reaction, but also for identification of this reaction as the penicillin-sensitive step in cell wall synthesis. Direct demonstration of transpeptidation *in vitro* was achieved originally in cell envelope

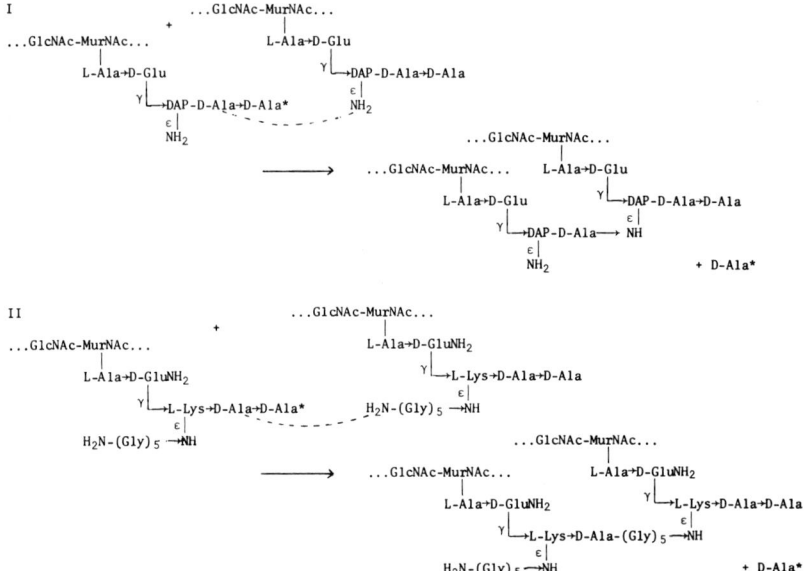

Figure 86 The bridge closure reaction in *E. coli* (I) or *S. aureus* (II). The reaction results in the liberation of one molecule of D-alanine (starred) and and the formation of the cross bridge between the resulting terminal D-alanine of one tetrapeptide side chain, and a free amino group in the other. In both I and II, the donor peptide is on the left left, the acceptor peptide on the right.

fractions from *E. coli*. Recently, D. Mirelman in our laboratory has demonstrated this reaction with crude cell wall preparations of *S. aureus* and *M. luteus* (A1,A2).

The evidence available clearly shows that crosslinking occurs according to the reaction described in Fig. 86.

In all cases investigated, incorporation of MurNAc-pentapeptide into peptidoglycan was accompanied by liberation of free D-alanine. The ratio of D-alanine to D-glutamate in the polymeric product approached unity and the expected crosslinked disaccharide-peptide dimer could be isolated following digestion of the enzymic product with lysozyme. The transpeptidation reaction, as measured by the release of D-alanine or the formation of crosslinks, was specifically and irreversibly inhibited by penicillin. This inhibition was paralleled by the appearance of an uncrosslinked product which retained the terminal D-alanyl-D-alanine structure of the precursor.

ROLE OF THE TRANSPEPTIDASE IN PEPTIDOGLYCAN BIOSYNTHESIS

The use of crude cell wall as enzyme source for peptidoglycan biosynthesis, introduced by Mirelman in 1972 (A1,A2), offered distinct advantages over the membrane preparations which served earlier for this purpose. The crude cell walls, which contain membrane fragments, were found not only to catalyze polymerization of the sugar nucleotide precursors UDP-GlcNAc and UDP-MurNAc-pentapeptide, but also to attach the newly synthesized peptidoglycan to the preexisting one in the cell wall. Apparently, such attachment does occur, because the enzymes involved, which are membrane bound, are intimately associated with the growing region of the preexisting peptidoglycan.

The detailed studies of Mirelman on *M. luteus* have shown that with crude cell wall preparations from this organism, incorporation of the nucleotide precursors into preformed peptidoglycan was accompanied by the release of *C*-terminal D-alanine from the MurNAc-pentapeptide. As expected, this release was completely inhibited by very low concentrations (less than 1 µg/ml) of penicillin. Very surprisingly, however, penicillin also inhibited markedly (up to 75-80%) the incorporation of *N*-acetylglucosamine and of MurNAc-pentapeptide into the preexisting peptidoglycan. On the basis of these findings, a new role was suggested by us (A1,A2) for the penicillin-sensitive transpeptidation reaction. We postulated that transpeptidation effects not only the formation of interpeptide crosslinks, subsequent to the transfer by transglycosylation of disaccharide units or linear peptidoglycan strands to the preexisting wall, but also, and more importantly, leads to covalent attachment of new strands to older ones (Fig.87). It thus appears that elongation and growth of the preexisting peptidoglycan is the result of two enzyme-catalyzed reactions: the penicillin-insensitive transglycosylation and the penicillin-sensitive transpeptidation.

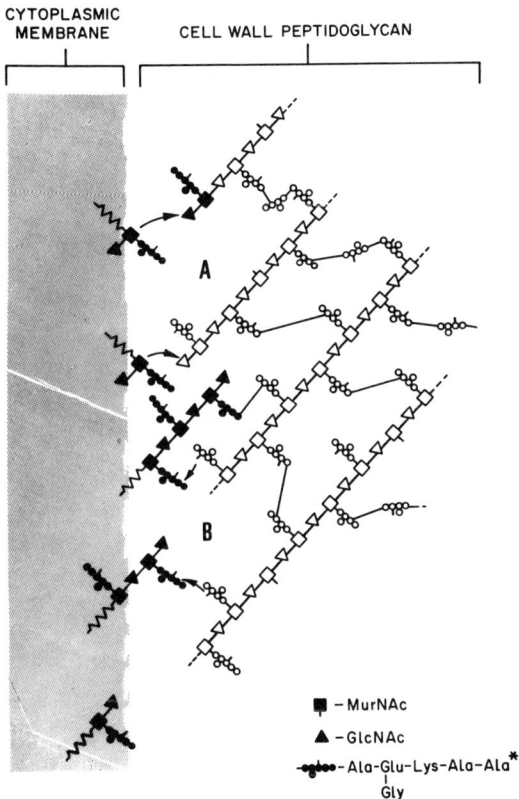

Figure 87 Model of growth of M. luteus cell wall peptidoglycan (A2). Newly synthesized strands are attached to preexisting cell wall peptidoglycan by two mechanisms. The main one is transpeptidation to an amino group on a preexisting peptide side chain with concomitant release of terminal D-alanine from the newly synthesized peptidoglycan chain (see B on figure). The second mechanism is the attachment by transglycosylation of an oligosaccharide-peptide intermediate to a non-reducing end of a preexisting glycan chain (see A). Preexisting peptidoglycan is shown in white and newly synthesized strands in black. The strands incorporated either singly or in polymerized form are depicted as lipid-bound although it is possible that the lipid moiety is removed before incorporation. (⋙ = glycosyl carrier lipid pyrophosphate).

The suggestion that the main role of the transpeptidase is to attach growing peptidoglycan strands to preexisting ones, explains a number of puzzling observations reported in the literature. It clarifies, for example, why low concentrations of penicillin inhibit the incorporation *in vivo* of cell wall amino acids such as L-lysine or glycine into the peptidoglycan of *S. aureus*. Furthermore, the dramatic killing effect of penicillin at very low concentrations may be more clearly understood if it affects attachment of newly synthesized peptidoglycan strands to existing ones, and not only formation of interpeptide bridges. A new role may also be envisaged for the cell wall endopeptidases that act, in many bacteria, on the peptide moiety of the cell wall peptidoglycan (see lecture 20, ref.2). One of the functions of these enzymes may be to prepare space for enlargement of the cell wall. For example, the liberation of the amino group of L-alanine (bound to the carboxyl of *N*-acetylmuramic acid) by an amidase in *M. luteus* may provide an acceptor moiety for the attachment of newly synthesized strands by transpeptidation resulting in the formation of D-alanyl-L-alanyl linkages. Such peptide crosslinks have indeed been found in *M. luteus* cell walls (Fig.77).

Further support for the role of the transpeptidase as

described above, has been recently obtained in a number of laboratories (see for example ref. A3), including ours (A4). Of special importance is the isolation by Mirelman of a linear peptidoglycan from the culture medium of M. *luteus* inhibited by penicillin. This uncrosslinked peptidoglycan, the peptide units of which contain the D-Ala-D-Ala *C*-terminal sequence, is most likely a macromolecular precursor in cell wall biosynthesis, the attachment of which to preexisting peptidoglycan has been blocked by penicillin.

TRANSPEPTIDASE AND DD-CARBOXYPEPTIDASE

Up to this point in our discussion we have encountered only one penicillin-sensitive enzyme, the transpeptidase. There is little doubt that among the numerous enzymes involved in peptidoglycan synthesis, this is the major target of action of penicillin and of related antibiotics, such as the cephalosporins.

In most bacteria, mainly in rod-like organisms, other penicillin-sensitive enzymes have, however, been identified. One of these is a D-alanine carboxypeptidase (DD-carboxypeptidase) that removes the *C*-terminal D-alanine from the D-Ala-D-Ala sequence. The precise function of this enzyme is unknown. It may catalyze removal of D-alanine residues from peptidoglycan strands, thus limiting the extent of crosslinking and of peptidoglycan synthesis.

In *E. coli* and *Bacillus subtilis*, the DD-carboxypeptidases are mostly membrane bound, whereas in *Streptomyces* they are found in the culture medium, and as a result can be purified and fully characterized. Extensive studies on the *Streptomyces* DD-carboxypeptidase have been carried out by J. M. Ghuysen and his coworkers at the University of Liège, in collaboration with H. R. Perkins in London. Based on the study of the mode of action of the enzyme on model peptide substrates, they concluded that the DD-carboxypeptidases are indeed transpeptidases that have undergone solubilization and have become uncoupled. The difference in the effective function of the enzyme could, at least partially, be a question of the availability of water. In an aqueous environment, nucleophilic attack by hydroxyl ions on the enzyme-peptide complex, after elimination of the terminal D-alanine, would lead to hydrolysis. In the hydrophobic environment of the membrane, attack by a suitable amine would lead to transpeptidation (R denotes peptide substituent on D-Ala-D-Ala, E denotes enzyme):

$$R\text{-}D\text{-}Ala\text{-}D\text{-}Ala + E \longrightarrow R\text{-}D\text{-}Ala\text{-}E + D\text{-}Ala$$
Carboxypeptidase activity

$$R\text{-}D\text{-}Ala\text{-}E + OH^- \longrightarrow R\text{-}D\text{-}Ala\text{-}COOH + E$$

Transpeptidase activity

$$R\text{-}\underline{D}\text{-Ala-}E + R'NH_2 \longrightarrow R\text{-}\underline{D}\text{-Ala-}NHR' + E$$

Serious doubt on the postulated identity of the DD-carboxypeptidase and transpeptidase stems from studies of these two activities in *E. coli*, *Bacillus subtilis* and related organisms, carried out mainly by Strominger and his coworkers. It was found, for example, that the DD-carboxypeptidase of *E. coli* is over 100 times more sensitive to inhibition by penicillin than the transpeptidase. The concentrations of penicillin required to inhibit the transpeptidase were very close to the concentrations required to inhibit growth of the organism, in accord with the assumption that the transpeptidase is the primary site of the lethal action of penicillin. Another important difference between the two activities in *E. coli* is that whereas inhibition by penicillin of the transpeptidase is irreversible, that of the carboxypeptidase is, in most cases, reversible and competitive. The situation is further complicated by the finding that *E. coli* contains several carboxypeptidases, and probably also more than one transpeptidase.

None of the bacterial membrane-bound transpeptidases have been isolated in soluble form and characterized, in contrast with the DD-carboxypeptidase activity that can be solubilized. The carboxypeptidase from *B. subtilis* has indeed been purified to homogeneity. This carboxypeptidase is irreversibly inactivated by penicillin, and the bound penicillin can be easily removed from the enzyme by treatment with hydroxylamine or with ethylmercaptan (ethanethiol). Penicilloyl hydroxamate or the ethylthio ester of penicilloic acid are formed, indicating that penicillin is bound to the enzyme as a penicilloyl derivative. Hydroxylamine not only removes the bound penicillin, but also restores the activity of the inhibited enzyme. It has been suggested that inactivation of the carboxypeptidase is the result of acylation of an essential sulfhydryl group on the enzyme by the highly reactive CO-N bond in the β-lactam ring of the penicillin molecule (Fig.88). This is supported by the finding that one of the four sulfhydryl groups of the enzyme disappears during the reaction with penicillin.

A mechanism of transpeptidation and its relationship to penicilloylation and DD-carboxypeptidase activity, as proposed by Strominger for *S. aureus*, is summarized in Fig. 88.

Examination of molecular models of penicillin, which is a cyclic dipeptide composed of two amino acids, D-valine and L-cysteine, led Tipper and Strominger to propose that the antibiotic is a structural analog of D-alanyl-D-alanine and

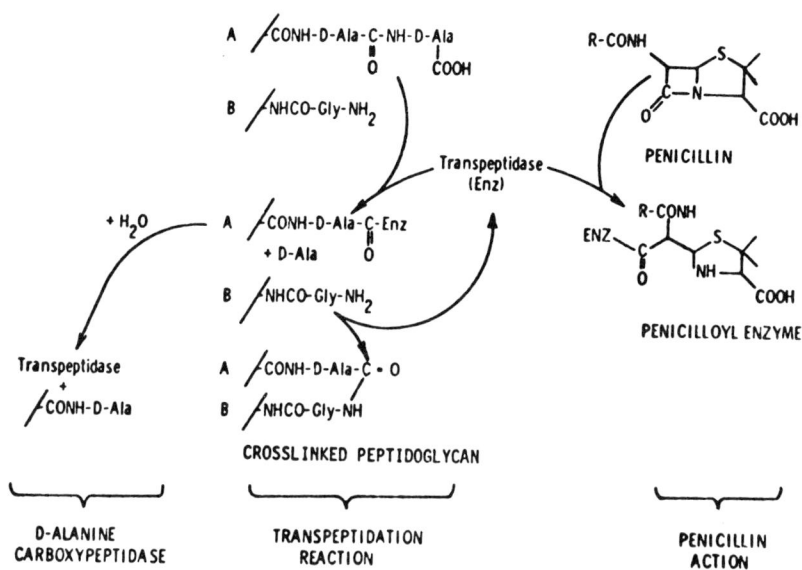

Figure 88 Proposed mechanism of transpeptidation and its relationship to penicilloylation and D-alanine carboxypeptidase activity (in S. aureus). "A" represents the end of the main peptide chain of the peptidoglycan strand. "B" represents the end of the pentaglycine substituent from an adjacent strand. If the acyl enzyme intermediate can react with water instead of the acceptor (left), the enzyme would be regenerated and the substrate released. The overall reaction would be hydrolysis of the terminal D-alanine residue of the substrate (DD-carboxypeptidase activity). (From ref.3).

can, therefore, readily interact with the enzyme. The CO-N- bond in the β-lactam ring of penicillin, which, as I just mentioned, is presumed to acylate the B. subtilis DD-carboxypeptidase, lies exactly in the same place in the model as the peptide bond in D-alanyl-D-alanine that is involved in the transpeptidation reaction. Because the corresponding peptide bond in the penicillin molecule is part of a four-membered ring structure (see Fig.88), it is somewhat strained. This is the basis of the suggestion that penicillin is in fact a transition-state analog of D-alanyl-D-alanine (A5); that is, an analog of the transition state involved in the breakage of the peptide bond in the transpeptidation (and possibly also of the carboxypeptidase) reaction.

ARCHITECTURAL RELATIONSHIPS IN PEPTIDOGLYCAN SYNTHESIS

At this juncture it is perhaps worthwhile to speculate on the relationship between the events in peptidoglycan synthesis and the molecular architecture of the cytoplasmic membrane and cell wall. The precursors of peptidoglycan (sugar nucleotides, aminoacyl-tRNA's, etc.) presumably originate in the cytoplasm, and it is reasonable to suppose that the assembly of the complete disaccharide-peptide repeating unit takes place at or near the inner cytoplasmic surface of the membrane. The peptidoglycan, however, lies outside the membrane and it seems probable that the final steps in synthesis, chain elongation and peptide crosslinking, occur at the outer surface of the membrane or possibly in the wall matrix itself. This implies, first, an asymmetric distribution of enzyme activities at the two surfaces of the membrane and, second, a translocation of disaccharide-(peptide)-PP-GCL from its site of synthesis at the inner surface to its site of utilization at the outer surface of the membrane. Indeed the possible transport function of the lipid coenzyme was emphasized in the initial studies of Strominger and his coworkers (1965), and the term "carrier lipid" is appropriate not only to the function of bactoprenol as intermediate carrier of the activated glycosyl group, but also to cross-membranal transport of the completed oligosaccharide-peptide repeating unit. The mechanism of the postulated translocation through the membrane and the question of whether additional membrane functions are required for the transport of GCL and its derivatives, are entirely unknown.

Another outstanding problem is that of the mode of cell wall growth, replication and regulation at the cellular level and in relation to the cell cycle. Further progress in this direction depends on better understanding the architecture of the membrane and of the "vectorial" enzymology that is involved in membrane transport. It also depends on technical improvements in the purification of the relevant particulate, synthetic and lytic enzymes, which in their state are membrane-bound, and upon the isolation of mutants which lack any of these enzymes.

REFERENCES

Reviews

1. Procaryotic cell division with respect to walls and membranes,
   M. L. Higgins and G. D. Shockman, CRC Critical Reviews in Microbiol. $\underline{1}$, 29-72 (1971).
   An interesting discussion of models of growth of coccus and rod shaped microorganisms, with special emphasis on peptidoglycan, based mainly on the work from the authors' laboratory. Includes also a brief and stimulating review of the role of autolytic enzymes in wall growth.

2. Assembly of bacterial cell walls,
   F. Fiedler and L. Glaser, Biochim. Biophys. Acta $\underline{300}$, 467-485 (1973).

3. Interaction of penicillin with the bacterial cell: penicillin-binding proteins and penicillin-sensitive enzymes,
   P. M. Blumberg and J. L. Strominger, Bacteriol. Revs. $\underline{38}$, 291-335 (1974).
   An up to date review from the laboratory which contributed perhaps most to our current views on this outstanding problem.

4. The role of membranes in the synthesis of macromolecules,
   M. J. Osborn, *in* Structure and Function of Biological Membranes (Ed. L. I. Rothfield), Academic Press, 1971, pp. 343-400.
   Excellent review, highly recommended.

5. Inhibitors of bacterial cell wall synthesis,
   E. F. Gale, E. Cundliffe, P. E. Reynolds, M. H. Richmond and M. J. Waring, *in* The Molecular Basis of Antibiotic Action, John Wiley, 1972, pp. 49-120.

Specific articles

A1. Role of the penicillin-sensitive transpeptidation reaction in attachment of newly synthesized peptidoglycan to cell walls of *Micrococcus luteus*,
D. Mirelman, R. Bracha and N. Sharon, Proc. Natl. Acad. Sci. USA 69, 3355-3359 (1972).

A2. Studies on the elongation of bacterial cell wall peptidoglycan and its inhibition by penicillin,
D. Mirelman, R. Bracha and N. Sharon, Ann. N.Y. Acad. Sci. 235, 326-347 (1974).

A3. Peptidoglycan synthesis in *Bacillus licheniformis*. The inhibition of cross-linking by benzylpenicillin and cephaloridine *in vivo* accompanied by the formation of soluble peptidoglycan,
Z. Tynecka and J. B. Ward, Biochem. J. 146, 253-267 (1975).

A4. Penicillin-induced secretion of a soluble, uncross-linked peptidoglycan by *Micrococcus luteus* cells,
D. Mirelman, R. Bracha and N. Sharon, Biochemistry 13, 5045-5053 (1974).

A5. Conformation of penicillin as a transition-state analog of the substrate peptidoglycan transpeptidase,
B. Lee, J. Mol. Biol. 61, 463-469 (1971).

INDEX

Abequose, 253,337-339,350,353-354,372, 374,376-377
ABO(H) blood group substances, 28,37, 186-187,336
   biosynthesis, 235-248
   chemistry, 215-234
   genetics, 235-240
   see also, blood groups
A_1A_2 blood groups, 244-245
5-Acetamido-3,5-dideoxy, L-arabino-2-heptulosonic acid, see NANA-7
2-Acetamidoglucal, 148
2-Acetamido-3-hydroxypropanol, 97
Acetylation, 79,149-150,294-295,381, 387
N-Acetylaminomannuronic acid, 328,396
Acetyl-CoA, 149,381,387
N-Acetylgalactosamine,
   in bacterial cell wall, 327,328
   in blood group substances, 28,220, 222-223,226,228-230,232,243-244, 246,252,255-256
   on cell surfaces, 211
   in glycolipids, 232
   in glycoproteins, 39-40,43,68-69, 76,111-112,147,162,164-165,252, 256

   in mucopolysaccharides, 258,263- 264,267,284,290,292-293,311,313
   oxidation by galactose oxidase, 112
N-Acetylgalactosamine-4-sulfate, 269, 290,292-293
N-Acetylgalactosamine-4-sulfatase, 311
N-Acetylgalactosamine-6-sulfate, 269, 290,292-293
N-Acetylgalactosaminitol, 228
α-N-Acetylgalactosaminidase, 83,223- 224
N-Acetylgalactosaminyl-serine (threonine), 68,74,111,162,230, 252,272,286-287
N-Acetylgalactosaminyltransferase, 162, 236,242-246,287,293
N-Acetylglucosamine, 4-5
   in N-acetyllactosamine, 23
   in bacterial cell wall, 16,323-324, 326-327,337,348-351,359,363, 396-406,408-411,418-450
   biosynthesis of, 129,169
   in blood group substances, 222, 226-230,236,241,243,246,248, 252,266
   in core oligosaccharide, 104-107

dolichol phosphate derivative, 171-173
in glycolipids, 230
in glycoproteins, 41,69-72,90-91, 102-107,122,124,127,158-160, 163,164,206
and lactose synthetase, 23
in lipid linked intermediates, 171-173
in mucopolysaccharides, 258,263-264,267,272-273,311
N-Acetylglucosamine-1-phosphate, 129, 169,171,423
N-Acetylglucosamine-6-phosphate, 129, 169
N-Acetylglucosaminidase, 25,89-91, 101, 105,311-312,400-402
N-Acetylglucosaminitol, 72,106
N-Acetylglucosaminyl-asparagine, 25, 68-72,87,90-91,105-110,122,124, 252,272
N-Acetylglucosaminyltransferase, 158
Acetyl groups
in LPS, 354,386-387
in mucopolysaccharides, 258,274, 275
in peptidoglycan, 398,400-403
N-Acetylhexosamine, 59
see also, individual compounds
N-Acetylhexosaminidase, 111,436
N-Acetylactosamine, 23,160,163
sequence in glycoproteins, 160, 163,164,230
N-Acetylmannosamine
biosynthesis of, 129,147-148,169
and sialic acid, 146-148,169
N-Acetylamannosamine-6-phosphate, 129, 148-149,169,350
N-Acetylmuramic acid, 16,395-411,414
N-Acetylmuramidase, 399-400
see also, lysozyme
N-Acetylmuramyl-L-alanine amidase, 399-400,405,408,410
N-Acetylmuramyl-peptide nucleotides
see UDP-MurNAc-peptides
N-Acetylneuraminic acid,
analysis, 58,338
biosynthesis, 129,146-153,169
on cell surfaces, 211
chemistry, 54,144-146
distribution in nature, 17,142-144

in glycoproteins, 40,104,111, 158-160,165,169,182
and KDO, 350
see also, sialic acids
N-Acetylneuraminidase
see neuraminidase
N-Acetylneuraminate-9-synthetase, 147
α_1-Acid glycoprotein, 34-35,48,67,103, 109,157,163,182,183,199,202, 206
Acid hydrolysis, 92-97
of glycoproteins, 53-56,58-59,61-62
of glycosides, 56-62,71-72,79,94
of mucopolysaccharides, 61-62
Acid reversion, 55-57
Acylneuraminic acids, 5,58,178,180
see also, sialic acids, individual compounds
Adhesion, intercellular, 189-190
ADP-Glc, 130
Affinity chromatography, 50,205,250
Agglutination, 37,189,216
see also, hemagglutination
D-Alanine, 323-325,396,422,439-440
release by transpeptidase, 439-440
DL-Alanine, 404
from β-elimination, 76
L-Alanine, 423-425
Alanine racemase, 425
D-Alanine peptides, 401,405-414
D-Alanyl-D-alanine, 424,425,437
D-Alanyl-L-alanine, 407-408
D-Alanyl-D-alanine synthetase, 425
L-Alanyl-tRNA, 433
Alditols, 62
Alginic acid, 131,295-296
Alkali
action on sugars, 72-73
cleavage of glycosidic linkages, 41,72-78,230,252,261,271-272, 285
see also, β-elimination
Amidase
see N-acetylmuramyl-L-alanine amidase
D-Amino acids, 394,400-415
Amino acids, linked to carbohydrates, 65-83
2-Aminoacrylic acid, 75,77
Aminoacyl-tRNA, 435
2-Aminobutyric acid, 76
2-Aminocrotonic acid, 75
3-Amino-3-deoxyglucose, 14-15
6-Amino-6-deoxyglucose, 15

3-Amino-3-deoxyribose, 14-15
6-Amino-6,8-dideoxy-D-erythro-
  D-galactooctose, 15
2-Amino-2,6-dideoxy-D-galactose, 340
2-Amino-2,6-dideoxy-L-galactose, 340
4-Amino-4,6-dideoxygalactose,
  340
2-Amino-2,6-dideoxyglucose, 340
4-Amino-4,6-dideoxyglucose, 340
4-Amino-4,6-dideoxyhexoses, 133
2-Amino-2,6-dideoxy-L-mannose, 340
Amino sugars, 13-16,340,394
  see also, hexosamines, diamino
    sugars, individual compounds
Analysis of sugars, 52-56,62-63,338
Amylase, 103-105
  see also, glucamylase
Antibiotics, 13-15,329
  see also, individual compounds
Antibodies, 208
  anti A, 216-217
  anti B, 216-217
  anti MN, 216
  anti type XIV polysaccharide,
    224,230,231
  production by lymphocytes, 209
Antifreeze glycoproteins, 39-40,68,
  112
O-Antigens
  see O-specific side chains
Anti O(H) reagents, 216
Antigen carrier lipid,
  see glycosyl carrier lipid
L-Arabinose, 43
D-Arabinose-5-phosphate, 349-350
L-Arabinosyl-hydroxyproline, 69,79,80
Arthritis, 260
Arthrobacter cyrstallopoietis, 433
Arylsulfatase B, 312
Ascarylose, 339
Asialo α_1-acid glycoprotein, 196,198-
  199,204
Asialoceruloplasmin, 194-196,199,204,
Asialofetuin, 196
Asialoglycopeptides, 199
Asialoglycoproteins, 194-207
Ashwell, G., 29,193,196,198,200,203,
  207
Asn-X-Ser(Thr) sequence, 108-111
Asparaginyl-N-acetylglucosamine,
  see N-Acetylglucosaminylasparagine
Asparaginyl carbohydrate, 86,90,99-
  102,105
Aspergillus niger, 181

ATP, 294,424-425,431,437
Autolytic enzymes, 402
  see also, autolysis
Autolysis, bacterial, 321-322, 402,430
  436
Avery, O.T., 26,326
Avidin, 109
Azotobacter vinelandii, 19,296

Bacillosamine
  see, 2,4-diamino-2,4,6-trideoxy-
    glucose.
Bacillus licheniformis, 18,326,328,
  340,435
Bacillus subtilis, 18,340,444,447
Bacitracin, 329,373,421,428,435-436
Bacterial cell wall, 10,16,22,318-
  342.
  see also, lipopolysaccharides,
    peptidoglycan
Bacterial mutants, 336,341,345-352,
  357,358,360
Bacterial transformation, 326-327
Bacteriophage
  conversion of O-antigens by,
    384-389
  ϵ^{15}, 385-389
  ϵ^{34}, 386-389
  lysogenic, 385-386,388-389
  receptors for, 16,321,323,329-332,
    384-386,389
  T-even, 20,110,329
Bactoprenol, 21,419,447
  see also, glycosyl carrier lipid
Bactoprenyl pyrophosphatase, 435-436
Bactoprenyl pyrophosphate
  see glycosyl carrier lipid pyro-
    phosphate
Baddiley, J., 16,322,325
Basement membrane, 67-69,78,113-115,
  156-157,187
Beadle, G.W., 239
Blix, G., 142
Blood group antigens
  see blood group substances
Blood group genes
  see blood group substances,
    genetics
Blood groups
  A_1A_2, 244-245
  ABO, 215-248
  "Bombay", 219,241,243
  Lewis (Le[a], Le[b]), 215,217,219-248
  MN, 216,218,248-256

Rh, 215-216
Blood group substances, 28,37,186-187,331
  biosynthesis, 236-248,256
  chemical structure, 28,215-233,248-256
  enzymatic degradation, 28,220,,222-224
  genetics, 217,235-248,256
  oligosaccharides from, 224,226-228
  in secretions, 37,221-222,233,237-238
  Se se gene, relation to, 235-238,242,245,247
Blood vessels, 260,263
Bogdanov, V.P., 70
Boivin, A., 333
"Bombay" blood group
  see blood groups
Borohydride, 72-76,145,195,201,209, 227-228,252,285,351-352,357
Borotritiate
  see borohydride
Braun, V., 330
Bromelain, 34,35,105
Brundish, D.E., 325
Burger, M., 29
Butler, W.T., 78

Calcium, 204-206, 261
Capsules, bacterial, 326,327
Carbohydrate-peptide linkages, 65-82,87,90,261,271-273
  see also, alkali, β-elimination, individual compounds
Carboxyl group, 58,77,179,259,360, 409-410
DD carboxypeptidase, 443-447
Cartilage, 284,286,272,278
Cathepsins, 379
CDP-Abe, 348,369,374,376,377
CDP-vinelose, 19,20
Celesticetin, 15
Cell adhesion, 189-190
Cell communication, 189-190
Cell cultures, 188,307-313
Cell envelope, 328,331,359-364,381
Cell membrane
  see membrane
Cellobiose, 9
Cellobiuronic acid, 27
Cellulose, 3,5,9,168

Cell surface, 17,22,27-29,160,178, 186-190,199-200,207-212,215, 218,263,329
Cell wall
  see bacterial cell wall
Cell wall disaccharide, 396-400,403-407
Cell wall tetrasaccharide, 397-399, 404,407
Centrifugation, density gradient, 49, 261,364-365
Cephalosporins, 443
Ceppellini, R., 236,247
Ceruloplasmin, 35,42,194-199,202,206, 316
Cervical mucus, 41, 180
Cetylpyridinium chloride, 262
Chain growth, direction of, 378-380
Chain termination 163,290
Chatterjee, A.N., 432
Chemotype
  of lipopolysaccharide, 341,349
  of peptidoglycan, 411-414
Chitin, 4,404
Choline phosphate, 325
Chondroitin sulfate, 40,262-264,267-272,278-279,283-295,298-300, 311,313
Chondroitin-4-sulfate, -6-sulfate
  see chondroitin sulfate
Chondroitinase, 286
Chondrosine, 264
Chondrosites, 283
Cifonelli, J.A., 273
Clamp,J., 7
CMP-KDO, 356-357
CMP-NANA, 125-126,129,146,150-153,158, 168-169,198,205
Colitose, 339
Collagen, 36,44,67-68,78-79,81,110, 113-115,155-157,184-185,259
Collagenase, 114
Colominic acid, 17,143,147
Colorimetry, 51-52,63
Competition experiments (enzymatic) 288,291
Concanavalin A (ConA), 44,49,50,208, 211,333
Conformation, 55
Connective tissue, 40,259-261,273, 279,306,310
Contact inhibition, 178,188-190

Core region
  of glycoproteins, 104-107,173-174
  of lipopolysaccharide, 341-342, 346-352,356-363,369,381-382
Cornea, 272
Corrective factors, 310,312
Crosslinkages, in peptidoglycan, 405-415,418-420,436-448
C substance, 325
Cunningham, L.W., 70,78,99
Cycloheximide, 282
D-Cycloserine, 421-422,425
Cysteic acid, 76
L-Cysteinyl-tRNA, 433
Cytoplasmic membrane
  see membrane

Dark, F.A., 16
Dehydroalanine,
  see 2-aminoacrylic acid
3-Deoxy-mannooctulosonic acid,
  see KDO
Deoxyribonuclease, 36,103
Deoxysugars, 58,131-138
  see also, individual compounds
Dermatan sulfate, 262-263,267,272, 276,279,295,297,305-308,311, 313
Detergents, 51-52,161,330
Diabetes, 113,115
Diacetylneuraminic acid, 17,144
Diacetylmuramic acid, 403
Dialdehyde, 93-94
Diamino acids, 331,412,414
2,3-Diamino-2,3-dideoxyglucose, 340
2,6-Diamino-2,6-dideoxyglucose, 14
2,6-Diamino-2,6-dideoxy-L-idose, 14
2,4-Diamino-2,4,6-trideoxyglucose, 18 325-326
Diaminopimelic acid,331,412-413,420
3,6-Dideoxy sugars, 133,339-340,369
  see also, individual compounds
Dinitrophenylation, 432
Diplococcus pneumonia, 325
Disaccharidases, 36
Disaccharide pentapeptide, from
  M. luteus, 405-406
Disaccharide peptides, 409,418,
  assembly of, 426-435
  polymerization of, 430,435-436, 440
Disaccharides, 8-9,111,165
  see also, individual compounds

Disaccharide units, 66, 401
  in glycoproteins, 39,111-113,156-157,165,187
  in mucopolysaccharides, 40,258,267, 269,272,289-291,293
  in peptidoglycan, see cell wall disaccharide, disaccharide peptides.

Dische, Z., 53
DNA, 156,163,327
Dolichol, 170
Dolichol phosphate, 21,173,371
Dolichol phosphate sugars, 21,170-174
Dopamine β-hydroxylase, 50
Dorfman, A., 273,289,303
Dysentery, 332

Edman degradation, 432
Electron microscopy, 278,319,328,331, 364,384
Eliminases, 267
Elimination reactions, enzymic, 135, 148,267
β-Elimination reaction, 74-77,81,111-112,228
  see also, alkali, cleavage of carbohydrate-peptide bonds
Elson-Morgan reaction, 63
Endopeptidases
  see peptidases
Endoplasmic reticulum, 121-128,283, 291-292,298
Endotoxins, 332,334
Enzyme replacement therapy, 315
Epimerization,
  chemical, 73,146
  enzymic, 73,129-131,138-140,147-148,295-298,404
Epimerase
  see epimerization, enzymic,
  UDP-Gal-4 epimerase,
  UDP-GlcNAc-2 epimerase
Erythrocytes, 28,34,69,81,199-200,215-216,219-221,224,231-233,237-240, 248-256
Erythropoietin, 34,201
Erythritol, 95-97
Escherichia coli, 18,20-21,110,133, 135,143,147,329,334,338-339,346, 356-357,394,413,419,428,438-439, 444-445.
Ethanolamine, 337,350
Exotoxins, 332

Extensin, 34,45,80
Eye, 41, 179, 260,263
Eylar, E., 183-184

Feed back control, 168-169,299-300, 390
Fetuin, 34,45,48,51,67,69,103-104, 107,157,163,186,195-196,203
Fibrinogen, 35
Fibroblasts, 307-310,312-314
Ficaprenol pyrophosphate galactose, 375
Fischer, E., 11
Fleming, A., 322
Fletcher, H. G., 147
Flowers, H.M., 397
Follicle stimulating hormone, 36,200-202
Food, carbohydrates as, 5
Food poisoning, 332,334
Fructopyranosides, 58
Fructose, 73,347
Fructose-6-phosphate,129,168-169, 299,347
Fucosamine,
   see 2-amino-2,6-dideoxygalactose
L-Fucose,
   analysis of 52,53,59
   biosynthesis, 131-138
   in blood group substances, 37,220, 222-224,226-229,236-237,242-243, 245
   on cell surfaces, 29, 188
   in glycoproteins, 37,69,103-104,109, 114,119-127,129,143,163-165, 189
   in milk oligosaccharides, 143
   in mucopolysaccharides, 262,264, 272
L-Fucosidase, 25,224
Fucosidosis, 25
2'-Fucosyllactose, 225,244
3'-Fucosyllactose, 225
Fucosyltransferase, 160,164,236-242, 246-247
Fukasawa, T., 346

Galactomuramic acid, 404
Galactosamine, 13,54,63,228,258,265, 270
   see also , N-acetylgalactosamine
Galactosamine-6-phosphate, 325
Galactose,
   analysis of 54,63
   in blood group substances, 28,37, 222-233,236-237,242-245,255-256
   on cell surfaces, 202,210-212
   in glycoproteins, 37,43,78,91,103, 109,112,114,119,121-129,155-156, 160,164-166,194-199,204,206-207, 316.
   in lipopolysaccharides, 337,341, 346-349,359-363,369,377,384, 386-388
   metabolism of, 23-24,139,347
   in mucopolysaccharides, 262-264, 271-272,284,287-289,291
Galactosemia, 24-25
Galactose oxidase, 63,112,194,197,209-211,360-361
Galactose-1-phosphate, 129
α-Galactosidase,25,28,83,223-224,
β-Galactosidase,89,91,197
Galactosyl-hydroxylysine,68,78-79,155-157,162,187
Galactosyl-serine,68
Galactosyltransferase, in biosynthesis, blood group substances, 236,245
   glycoproteins, 156-158,160,162,166
   lipopolysaccharides, 362-366
   mucopolysaccharides, 287-288,291-292,372-373,376
Galacturonic acid, 360
Gal-β-(1→3)-GalNAcα(1→)Thr, 39,112
Gal-β-(1→4) GlcNAc
   see N-acetyllactosamine
Gal-β-(1→4) Xyl, 287-288
Gal-1-P-uridyltransferase, 24-25,347
Gal-PP-GCL, 372-373,376
Gangliosides, 25,142-143,159
Gas liquid chromatography, 62,63,76, 92,265
Gastrointestinal,
   see intestinal
Gaucher's disease, 316
GDP-Glc, 168
GDP-L-Fuc, 125-126,129,132-134,138, 240-241
GDP-4-keto-6-deoxymannose, 132-133
GDP-Man, 129,131-134,138,166,172,369, 374
Gel electrophoresis, 51-53,85,251
Gel filtration, 51,85
Genetic control, 110,388
   see also, blood group substances, lipopolysaccharides
Genetic defects, 24-25,42,303-317
   see also , bacterial mutants
Genetic diseases,
   see genetic defects

[458]

Gene interaction, 247-248
Gentian violet, 422
Gentiobiose, 8
Gesner, B. M., 29
Ghalambor, M. A., 338
Ghuysen, J. M., 396,412,444
Ginsburg, V., 29,239
GlcNAc-β-(1→4)-MurNAc
   see, cell wall disaccharide
GlcNAc-β-(1→4)GlcNAc-β(1→)Asn, 106
   see also, core region of glycoproteins
GlcNAc-MurNAc-(peptapeptide)-GCL, 430-432
Glick, M. C., 189
Glucamylase, 68,181
Glucosamine, 13,265
   analysis of, 61-63
   in glycoproteins, 114,119,122-123,169,188,232
   in lipopolysaccharides, 334,337,341,350,355
   in mucopolysaccharides, 258-259,265-266,272-275
   in peptidoglycan, 394
   see also, N-acetylglucosamine
Glucosamine-6-phosphate, 129,168-169
Glucosaminides, 59-61
Glucosaminitol, 105,402
Glucose,
   alkaline degradation of 73,
   in bacterial cell wall, 323-328,396
   in glycolipids, 160,233
   in glycoproteins, 58,110,114,156,171
   in lipopolysaccharides, 337,341,346-349,353,358-363,375,377-378,383,386-388.
   metabolism of 129-131,139,347.
   in milk oligosaccharides, 143.
   oligosaccharides of, 8,20,23,27
   in polysaccharides 4-5,20,27,96
Glucose-1-phosphate,129,347
Glucose oxidase, 45,50,63
β-Glucosylceramide, 316
Glucosylceramide β-glucosidase, 315
Glucosylgalactosyl hydroxylysine, 79, 113-115,187.
Glucosyltransferase,in biosynthesis
   of collagen, 156-157
   of lipopolysaccharides, 360-366, 388.
Glucuronic acid, 27,44,131,258,263,

269-276,284,289-290,296-297,313,328
β-Glucuronidase, 311-313
Glucuronyl-β-(1→3)galactosamine
   see chondrosine
Glucuronyl-β-(1→3)glucosamine,
   see hyalobiuronic acid
Δ4,5-Glucuronyl-β-(1→3)N-acetylglucosamine, 266-267
Glucuronyltransferase, 289-293
D-Glutamic acid,396,404-406,414-415,424,426,428,430-431
Glutathione, 431
Glycerol, 94-97,323,334
L-Glycero-D-mannoheptose,
   see heptose
Glycine, in peptidoglycan, 396,404-406,409-413,431-434.
Glycogen,4,20,22,25,44,130,168
Glycolipids,17,25,130,158,188,202,231-232,238-239
   see also, gangliosides, phosphatidylethanolamine
N-Glycolylmuramic acid, 403
N-Glycolyneuraminic acid, 17,144,149-150
Glycopeptides,
   from glycoproteins, 84-91,98-115,189,199,203,249,252-256
   from peptidoglycan, 404-406,409
Glycophorin, 248-255
Glycoproteins,
   analysis, 44-45,52-63
   biosynthesis, 21-22,118-128 155-175
   chemical structure, 65-69,86-87,99-115
   distribution in nature, 33,38-42,45,79-80
   function of carbohydrate in,177-188,193-207
   isolation,48-50
   microheterogeneity, 85,99-103,156,162,231,248
   physicochemical properties, 48,51-52,178-182
   sugar constituents, 42-44
   see also, blood group substances, carbohydrate-peptide linkages, glycopeptides, mucopolysaccharides, individual compounds
Glycosaminoglycans,
   see mucopolysaccharides
Glycosidases,28,87-91,223,279,352,396

[459]

see also, individual enzymes
N-Glycosidic linkages, 67-69,87,
122,212
O-Glycosidic linkages, 68-69,73-74,
78-81,87,113,181,251,271
Glycosides, hydrolysis of,
see acid hydrolysis, alkali,
Glycosphingolipids, 221,230,238
Glycosyltransferases, 125,156-165,190
236-248,286-292
see also, individual enzymes
Glycosyl carrier lipid, 21,298,369-
382,419,427
see also, Gal-PP-GCL,bactoprenol,
bactoprenyl derivatives, dol-
ichol
Glycosyl carrier lipid phosphate
sugars, 21-23,369-382,388,419-
420,426-433,435-436,441, 449
see also, dolichol phosphate
sugars
Glycyl-tRNA, 432
Goebel, W.F., 27,326
Golgi apparatus, 120-128,153,163,283,
291-292
Golgi, C., 120
Gottschalk, A., 28
Gram-negative bacteria, 320,328-342,
394.
Gram-positive bacteria, 320,322-324,
340,394
Gram stain, 319
Ground substance, 259
Group specific antigens, bacterial,
327
L-Guluronic acid, 131,296-297

Hakomori, S.I.,91,221,231
Half-chair conformation,57
Hall, C.W., 299
Hansenula wingei, 37
Hapten inhibition, 224-226,243,352
Haptoglobin,65,195
Hare, R., 27
Harpaz, N., 28
Haskall, V., 278
Haug, A., 296
Haworth, W.N.,10
H blood type
see blood groups, blood group sub-
stances
Heath, E.C., 170,338
Heidelberger, M., 28,326
Hela cells, 124

Hemagglutination, 28,183-184,206,
224
Hen egg-white, 36,65
Heparan sulfamidase,25,311-313
Heparan sulfate, 262-263,272,275,279,
294,305,308,311-313.
Heparin,40,131,262-263,272-276,294-
295,297,313
Hepatic binding glycoprotein, 205-
206
Heptose, 337-341,348-352,358-363
Hers, G.H., 306
Hexosamine,37,54,114
see also, individual compounds
Hexokinase, 134
Hexosaminidase, 25,279
see also N-acetylgalactosaminidase
N-acetylglucosaminidase, N-ace-
tylhexosaminidase
Hirs, C.H.W., 101
Hirst, G.K.,27
Histocompatability antigens, 186
Horecker, B.L.,359-360,364
Hormones, 34,36,199-202
see also, individual compounds
Hudson, C.S., 10
Human chorionic gonadotropin, 34,65,
69,200,202
Hunter's syndrome,25,303-313
Hurler's syndrome,25,303-313
Hyaluronidase,266-267,270,276,279,
286
Hyalobiuronic acid, 264,266
Hyaluronic acid, 262-268,278,279,290,
327-328
Hydrazinolysis, 405
Hydroxylation, 66,149-150,156
Hydroxylysine, 66,113-114,156,184,187
Hydroxyproline, 66,79-80,114
5-Hydroxymethylcytosine, 20,110
3-Hydroxymyristic acid, 334,355-357

Iduronate sulfatase,25,311-312
L-Iduronic acid, 43-44,131,258,263,
272-276,295-297,313
α-L-Iduronidase,25,311-312
Immunoglobulins, 34-36,67-69,107,109,
127-128,186-188,333
Immunological determinants, 16,28,186-
187
see also,blood group substances,
lipopolysaccharides
Inbar, M., 29
Influenza virus, 28,183,249,253,255

Inner core,
  see core
Inner membrane, 384
Interferon, 38, 182
Intercellular adhesion
  see cell adhesion
Intestinal tract, 41, 179, 334
Intrinsic factor, 182
Invertase
  see yeast invertase
Ion exchange chromatography
3,5-Isomerase, 137

Jeanloz, R.W., 18, 170, 265, 375, 397

Kabat, E.A. 27, 221, 226-227
Kanamycins, 14
Katchalski, E., 207
Kathan, R.H., 251
Kauffman, F., 335
Kauffmann-White scheme, 335, 341
KDO, 337-339, 347, -350, 356-359
KDO-8-phosphate synthetase, 349-350
2-Keto-3-deoxyoctanoic acid
  see KDO
Keratan sulfate, 40, 262-264, 272, 273, 278-279, 286-287, 306, 311
Killer cells
  see lymphocytes, cytotoxicity
Klenk, E., 142
Kojibiose, 323
Kornfeld, S., 203
Koschielak, J., 231
Koshland, D.E., 167
Kuhn, R., 224

α-Lactalbumin, 23-24
Lactobacillus, 324-325, 327
Lacto-difucotetraose, 225, 243
Lacto-N-difucohexaose I, 225-226, 243
Lacto-N-fucopentaose I-II, 225-226, 244
Lactose, 20, 23, 24, 199, 224
Lactose synthetase, 23-24, 287
LaForge, L.F., 13
Landsteiner, K., 216, 249
Larsen, B., 296
Le^{a}Le^b, blood groups,
  see blood groups, blood group substances
Lectins, 206
  blood group specific, 216-217, 220, 232, 249
  interaction with cells, 29, 178, 189, 206, 333, 253, 389

for isolation of glycoproteins, 49-50
  see also, individual compounds
Ledderhose, G., 13
Leloir, L.F., 10, 19, 128, 170, 395
Levene, P.A., 13
Leucocytes, 315
Lewis blood groups
  see blood groups, blood group substances.
Lincomycin, 15
Lindahl, U., 297
Lindberg, B., 92
Linkage region,
  in glycoproteins,
    see core region
  in mucopolysaccharides, 270-272, 283-289, 293
Lipid A, 334, 337, 340-342, 349-350, 355-359, 376, 383
Lipid linked sugar intermediates
  see dolichol phosphate sugars
  glycosyl carrier lipid phosphate sugars.
Lipidoses, 315
Lipmann, F., 293
Lipopolysaccharides,
  biological properties 329-336, 384-385
  biosynthesis, 21-22, 345-364, 369-390.
  chemical structure, 336-342, 345-363.
  isolation of, 330
Lipoproteins, 320, 330-331
Lipoteichoic acid, 371
Liver, 122, 124-125, 163, 193-200, 203-207, 305-307
Lobry de Bruyn-transformation, 73
Lüderitz, O., 355
Luteinizing hormone, 36, 201
Lymphocytes,
  cytotoxicity, 212
  homing, 28-29, 178, 188
  mitogenic stimulation, 187-188, 207-212, 333
Lysogenic infection,
  see bacteriophages, lysogenic
Lysosomal diseases, 307, 315
Lysosomes, 126, 194, 197, 203, 306-307, 312, 379.
Lysostaphin, 321, 400-401, 410
Lysozyme, 7, 10, 23, 35, 36, 44, 168, 321, 329, 396-407, 410, 439

Lytic enzymes
see autolysis, individual enzymes.

MacLeod, C.M., 327
Malignant transformation, 29, 188-189, 389
Man-β-(1→4)-GlcNAc-β-(1→4) GlcNAcβ-(1→)-Asn, 106-107
see also, core region of glycoproteins.
Maltose, 9, 36
Mannan, 371
see also, yeast
Mannomuramic acid, 403-404
Mannosamine, 142, 146
see also, N-acetylmannosamine
Mannose, 54-55, 73
in glycoproteins, 43, 69, 81, 88, 99-107, 109, 114, 119-123, 127-128, 166, 171-182, 188, 206
in lipopolysaccharides, 337, 351-354, 369, 374, 376-378, 383
metabolism of, 129, 131-134, 166
in mucopolysaccharides, 262, 264, 272.
α-Mannosidase, 88-90, 100, 105, 351
β-Mannosidase, 89-90, 105
Mannosidosis, 25
Mannosyl-serine, 68
Mannosyl-threonine, 68
Man-L-Rha-Gal-, 352-354, 371, 375-377, 386-388
Man-L-Rha-Gal-PP-GCL, 372-380
Mannuronic acid, 131, 296-297
Marchesi, V.T., 251
Maroteaux-Lamy syndrome, 305, 311-313.
Mass spectrometry, 12, 92, 370
Mast cells, 260, 263
Mastocytoma cells, 212, 285, 294
McCarty, M., 327
Melibiose, 351
Membrane, 68, 120, 126, 161, 174, 178, 180, 188, 204-206, 260, 319, 363, 366, 380, 383, 436, 439-440
see also, cell surface, outer membrane
Membrane transport intermediates,
see transport of lipid linked sugars.
Methylation analysis, 91-92, 106, 265, 350, 352
Methyl α-N-acetylglucosaminide, 59-60

N-Methyl-L-glucosamine, 13
Methyl α-glucoside, 50, 93
Methyl α-mannoside, 49
Meyer, K., 259, 262, 266
Micrococcus luteus, 322, 328, 371, 394, 396-400, 403-414, 419, 421, 430-431, 438-443.
Micrococcus lysodeikticus,
see Micrococcus luteus
Micrucuccus roseus, 433-434
Microheterogeneity,
see glycoproteins
Microsomes, 171
Milk oligosaccharides, 142-143, 224-225, 239-240, 245
see also, individual compounds
MN blood group,
see blood groups, blood groups substances.
Mirelman, D., 405, 438-440, 443
Mitogens, 208-210, 333
see also, lymphocyres, mitogenic stimulation.
Morell, A.G., 29, 193, 196, 198, 200, 203, 207.
Moggridge, R.C.G., 59
Montgomery, R., 100
Morgan-Elson reaction, 397
Morgan, W.T.J., 27, 220-221, 235, 247, 333
Morphogenesis, role of carbohydrates in, 188
Morquio syndrome, 305, 311
Mucins, 41, 68, 111, 178-180
see also, submaxillary mucins.
Mucopolysaccharides,
biosynthesis, 282-300
chemical structure, 40-42, 61-62, 77-78, 258-279
genetic defects, 24-25, 303-316
Mucopolysaccharidoses,
see mucopolysaccharides, genetic defects.
Muir, H., 278
Multienzyme complexes, 291-292, 295, 300
Multiglycosyltransferase system, 159-164
Muramic acid, 16, 394
see also, N-acetylmuramic acid.
Muramicitol, 402
Murein,
see peptidoglycan
MurNAc-β-(1→4) GlcNAc, 398, 401, 403
MurNAc-(pentapeptide)-PP-GCL, 429
Mutants,

[462]

see bacterial mutants.
Myelin, 38
Myeloma cells, 127

NAD$^+$,129,133-139
NADH,135-136
NADP$^+$,129
NADPH, 133-134,137-139,423
NANA-7,201-202
NANA-α-(2$\rightarrow$6)-GalNAc, 111
Nasal septum, 276-277
Neomycins, 14
Neuberger, A., 59,70,108
Neufeld, E.F.,299,303,308
Neuhaus, F.C., 419
Neuraminic acid,16-17,58,142,179-180, 183,338
see also, sialic acid,individual derivitives of neuraminic acid.
Neuraminidase, 28,87,89-91,111,183, 194-207,209-212,249
Nigeran, 96
Nikaido, H., 345,390
p-Nitrophenyl-β-D-xyloside, 283
Nonulose,18
Northcote, D.H.,10
Novogrodsky, A., 207-211
Nuclear magnetic resonance, 12,105, 265,274
Nucleoside monophosphate sugars,
see CMP-KDO, CMP-NANA
Nucleus pulposus, 263,272,278

Orosomucoid,
see α_1-acid glycoprotein
Osborn, M.J., 359-382
O-specific side chains, 329,333,335, 341-342,345,346,352-355,369-389
Outer membrane, 328-331,382-384
Ovalbumin,34-35,65,70-71,86-87,91,98-101,103,105,109,186
Ovarian cyst fluid, 221
Ovomucoid, 81
Oxidoreductases, 133-136

Pancreatic juice, 36,184
Paratose, 339
Park, J.T., 395,419,432,437
Paromomycins, 14
Pazur, J.H., 181
Penicillin, 22,321,329,395,420,437-447
Peptidases, 400-401,408,410,442

Peptide bridges,
see crosslinkages
Peptide units, (side chains)
see peptidoglycan
Peptidoglycan, 21,327-331,
biosynthesis,21,370,373,418-450
chemical structure,393-415
Perkins,H.R., 419,444
Periodate oxidation, 92-98,144-145, 181-182,187,201-202,209-212, 313,351-352,401
see also, Smith degradation
Phage,
see bacteriophage
Phelps, C.F.,35
Phillips, D.C., 10,23
Phosphate bridges, 350,359
Phosphatidylethanolamine, 365-366
Phosphodiester linkages, 323,350,355, 358-359,374.
see also, phosphate bridges.
Phosphoenolpyruvate, 147-149,349-350, 422-423
Phosphoglucomutase,134
Phosphoglucose isomerase,347-348
Phospholipids, 364-366
Phosphomannose isomerase, 383
Phosphonomycin, 422-423
Phosphorylation, in LPS, 358
3'Phosphoadenosyl-5'-phosphosulfate, 293-294,297.
Phytohemagglutinin,208,252
Phytotoxic glycopeptides, 68
Pinocytosis, 161,312
Plant cell walls, 68 80
Plant glycoproteins, 34,45,68,80
see also , bromelain, soybean agglutinin
Plasma glycoproteins, 33-36, 124, survival in circulation,193-207
Plasma membrane,
see membrane
Plasmacytoma cells, 122,127
Plummer, T.H.,101
Pneumonia, 325,326
Pneumococcal polysaccharides, 27,325-327
see also, Type XIV pneumococcus.
Polyacrylamide gel electrophoresis,
see gel electrophoresis
Polyisoprenoid compounds,
see bactoprenol,dolichol, glycosyl carrier lipid, and their derivitives.

Polymerization, 22, 289-293, 372-380, 386-388, 419-420, 435-436, 440-441.
Polymer
  modification of, 295-298
Porcine submaxillary mucin,
  see submaxillary mucins
Potato lectin, 45, 80
Prenatal diagnosis, 314
Proteoglycans,
  see mucopolysaccharides
Proteolytic digestion, 36-38, 84-85, 182, 249, 252, 261, 277, 285
Protoplasts, 393
Puromycin, 15, 121-122, 282-283
Pyrogenicity, 332
Pyrophosphatases, 132
  see also, bactoprenyl pyrophosphatase
Pyrophosphate bridges,
  see phosphosdiester linkages
Puruvic acid, 77

Quaternary ammonium salts, 49, 262.
Quinovosamine,
  see 2-amino-2,6-dideoxyglucose
Quinovose, 136.

Racemase,
  see alanine racemase
Radioautography, 119, 128
R-core,
  see core region, of lipopolysaccharide
Reductase, 137-138, 623
Rees, D.A. 20
Repeating units,
  see O-specific side chains
Respiratory tract, 41, 179
R-forms,
  see Salmonella
Rh,
  see blood groups
L-Rha-Gal-PP-GCL, 372-374
L-Rhamnose, 133, 136-137, 327, 369, 377
L-Rhamnosamine, 340
Richtmeyer, N.K., 18
Ribitol, 17
  see also, teichoic acids
Ribonuclease, 7, 44, 108, 124, 182, 432.
Ribonuclease B, 34, 36, 65, 67, 91, 101-103, 106-109, 181
Ribonuclease C and D, 36

Ribosomes, 119-125, 283
Ristocetin, 436
R-mutants,
  see Salmonella
Robbins, P.W., 293, 335, 352, 370, 378, 379, 386, 388
Roden, L., 126, 270
Rogers, H.J., 419
Rogers, J.C., 203
Roseman, S., 189
Rosenberg, L., 278
Rothfield, L.I., 359, 361, 366

Saliva, 41
Salmonella, 329-330, 334-336, 339-341, 346, 351-358, 376-377, 383
  R-forms(mutants), 341, 346-350, 351, 355, 381-382
  S-forms, 341, 346, 349, 355,
  P⁻mutants, 351, 358
  see also, individual organisms
Salmonella anatum, 352, 354, 381, 385-388
Salmonella minneapolis. 354
Salmonella newington, 354, 371
Salmonella typhimurium, 337, 341, 348, 350, 353-354, 359, 362-363, 371-374, 376, 383-384
Salo, W., 147
Salton, M.R.J., 319, 394, 396
Sanfilippo syndrome, 25, 304-305, 307, 310-312
Schachter, H., 126, 178, 244
Schauer, R., 150
Scheie syndrome, 310-312
Schwarz, U., 330
Secretion 124-128, 183-185
Secretors, 237-238, 241
Secretory granules, 123, 125-126
Sese gene,
  see blood group substances
Sephton, H.H., 18
Serum albumin, 33, 203
L-Seryl-tRNA, 434
Sexual agglutination
  see yeast
S-forms,
  see Salmonella
Sheep submaxillary mucin
  see submaxillary mucins
Shettles, L.B., 53
Shigella, 329
Sialic acids,
  analysis, 53-54, 338

biosynthesis,129,147-153,169
in blood group substances, 222
on cell surfaces 17,28,180-181, 202,207-212
chemistry, 144-146
distribution in nature,16-18,142-144,147-148
in glycoproteins,29,49,69,91,103-104,109,111,114,121,123,126-127,158,163-166,179-180,182-183, 193-207
in mucopolysaccharides, 264,272
and KDO, 338,358
see also, individual compounds
Sialic acid, 7-carbon analog,
see NANA-7
Sialyltransferases, 146,152,158-160, 163-166,198,205,207.
Silbert, J.E., 293-297
Smith degradation, 91-98,105-106,285
Smith, F., 92
Sodium dodecyl sulfate, 51-52,330
Sodium borohydride,
see borohydride
Soybean agglutinin, 85-86,90,103,182, 211,333
Spheroplasts, 393
Spiro,R.G.,67
Springer G.F,, 249,255-256
Staphylococcus aureus, 394-395,398-403,407,409-410,412,419-425,428-432,434-435,437-438,446.
Staphylococcus epidermis, 433-434.
Starch, 4,9,10,20
Staub, A.M., 335,353
Strange, R.E., 16
Streptococci, 324,327-328
Strptomycin, 14
Streptomyces, 444
Strominger, J.L.,10,170,370,395,414, 422,432.
Submaxillary mucins, 34,66-67,111,165
Sucrose, 5,20
Sugar nucleotides, 18-22,128-131,151, 158,166-170,173,190,283,294,345, 380,419-422,430.
see also, individual compounds
Sulfate, 40-41,49,258-263,265,273-276
Sulfated glycoproteins, 40-41
Sulfation, 282,292-295,297
Sulfotransferase, 293-297
Synovial fluids, 260,263

Tamm and Horsfall glycoprotein,183
Tatum, E., 240
Tay-Sachs disease, 25-26
TDP-Glc, 133-135
TDP-4-keto-6-deoxyglucose, 134-138
TDP-L-Rha, 133-138,369,372-373
Teichoic acids, 16,320,322-326,401
Teichuronic acid, 328
Teratoma cells, 188
Termination,
see chain termination
Thiobarbituric acid assay, 338
Thioglycosidic linkage, 81
Threitol, 106
L-Threonyl-tRNA,434
Thyroglobulin, 34,36,51,67,69,103,106, 109,124,186
Tipper, D.J., 437
Toxins, 34
see also, endotoxins, exotoxins
Transacetylase, 387-388
Trasferrin, 34-35,67,107,196,203
Transformed cells,
see malignant transformation
Transition-state analog, 447
Transglycosylation, 420,426-430,440-441
Translocation, 367,383-384,447, 449
Transpeptidase,
see transpeptidation
Transpeptidation, 420,437-447
Transport,
of glycoproteins,
see secretion
of lipid linked sugars, 21-22,371 430, 449
Trehalose, 8,20
Trichomonas foetus , 223
Trimethylsilyl sugars, 76
Tularemia, 332
Type 1 and Type 2 groupings, 227-230, 233,244,246
Type XIV pneumococcus, 224,230,231
Typhoid fever, 332
Tyvelose, 339

UDP,426
UDP-L-Ara, 140
UDP-Gal, 23,125-126,129-130,139-140, 156-158,168,245,246,286,287,293, 346-348,360,363,365,369,372-373.
UDP-Gal-4 epimerase, 138-140,346-347, 384

UDP-Glc, 20,21,128-130,139,156,168,
    172,299,347-348,358-359,363,
    365,388,395
UDP-GalNAc, 129,140,242,246,284,289,
    293,299
UDP-Glc dehydrogenase, 299
UDP-GlcNAc, 124-126,129,147-148,158,
    168,169,171-173,294,299,420-
    424,426,428,429,439
UDP-GlcNAc-2 epimerase, 169
UDP-GlcNAc-3-pyruvate enol ether, 423-
    424.
UDP-Glc pyrophosphorylase, 347
UDP-GlcUA, 129,284,289,293,294,297,
    299-300.
UDP-GlcUA, decarboxylase, 138
UDP-L-Iduronic acid, 295
UDP-MurNAc-L-Ala, 421,423,426
UDP-MurNAc, 420-424,429
UDP-MurNAc-peptides, 395,420-428,432,
    437,439
UDP-Xyl, 129,138,140,284-286,293,299-
    300
Uetake, H., 385
UMP, 372-373, 426,428
Urine, 239,305-306
Urogenital tract, 179
Uronic acids, 40,49,62,264
    see also, individual compounds

Vancomycin, 329,421,436
van Hoof, F., 306
Viruses 178,189,384-385,389
    see also, bacteriophage, influenza
    virus.

Warren, C.D., 375
Warren L., 189
Watkins, W.M., 220,221,235,239,247
Westphal, O., 27,330, 335,341,355
Wheat germ agglutinin, 250,254
Whelan, W. J., 10
White B., 335
Wilson's disease, 42
Winterburn, P.J., 35
Winzler, R.J., 251
Wolfrom, M.L., LL
Wright, A., 388

Xylitol, 285
Xylosaminitol, 105-106
Xylose, 43,138,262,271,283-285,300
Xylosyl-serine, 68,77,162,271,285,287,
    306.
Xylosyltransferase, 284-287,291-293

Yamakawa, T., 231
Yamashina, I., 70
Yeast
    cell wall mannaus, 68
    invertase, 68,81
    sex factors, 37-38
    sexual agglutination 37-38

Titles of Related Interest

Armin C. Braun
THE BIOLOGY OF CANCER

This book was written not for the specialist but rather for the beginner who has acquired a background in the basic biological sciences and who is interested in gaining insight into our present understanding of the cancer problem. The biology of cancer is examined here in the broadest terms, and an attempt is made not only to identify the essential biological concepts that underlie the tumorous state but to critically evaluate as well the premises upon which prevailing thought in the field of experimental oncology is based.

"Cancer research quite understandably has focused very heavily throughout the past century on immediate causes and cures. This clinical approach to the problem has been of limited effectiveness, producing 'treatments' but no widely applicable cures. It might appear desirable to gain an understanding of the basic cellular mechanisms that underlie the tumorous state . . . . While an attempt was made in this general survey to cover, albeit briefly, most if not all aspects of cancer biology, emphasis has been placed here on those investigations that appear most likely to provide insight into an understanding of that problem.

"This was done not only to stress the purpose and significance of pertinent studies that bear on these matters but, more important perhaps, to provide a conceptual framework for possible future investigations in this challenging and most important area of the scientific endeavor."— *From the Preface*

 1974, xii, 169 pp., illus.;
 hardbound ISBN 0-201-00764; paperbound ISBN 0-201-00765

CONTENTS:
Introduction · The Development of Autonomy · On the Origin of the Cancer Cell: Somatic Mutation · Addition of New Genetic Information · Epigenetic Changes · Biological Approaches to Control · References · Index.

Eric G. Ball
ENERGY METABOLISM

Written for the student or research worker who desires to obtain a coherent and concise knowledge of the basic reactions which are concerned in the conversion of foodstuffs into useful energy as well as the more recent developments in the pathways and hormonal controls which govern the storage and mobilization of carbohydrate and fat in the animal body.

 1973, xii, 84 pp., illus.
 hardbound ISBN 0-201-00406-2; paperbound ISBN 0-201-00407-0

CONTENTS:
Biological Oxidations and the Basic Pattern of the Process. Fatty Acid Oxidation of Acetyl CoA. Electron Transmitter System. Oxidative Phosphorylation. Cellular Localization of Reactions. Energy Storage. Energy Mobilization and Glucogenesis. Index. Glossary. References.

HORIZONS IN BIOCHEMISTRY AND BIOPHYSICS

E. QUAGLIARIELLO, Editor-in-Chief, and F. PALMIERI, Managing Editor
(University of Bari)
T.P. SINGER, Consulting Editor (University of California, San Francisco, School of Medicine)

Contributors to forthcoming volumes

Thomas B. Bradley, Jr.	A. Fonyo	H. L. Kornberg
Pierre Desnuelle	H. W. Heldt	G. Meissner
Harvey F. Fisher	P. J. F. Henderson	Paul Mueller
S. Fleischer	Ulrich Hopfer	J. Edwin Seegmiller

The aim of *Horizons in Biochemistry and Biophysics* is to call the attention of students, teachers, and practicing scientists in the biological and physical sciences, including medicine, to

* major conceptual and methodological advances and important discoveries in biochemistry and biophysics
* the need for re-evaluating widely accepted theories
* the possibility of applying discoveries to knowledge in other fields.

In order to ensure the wide readership vital to the concept of the undertaking, the initial volumes of *Horizons* will be issued simultaneously in hardbound and in paperbound form. It is hoped that the relatively modest price for the paperbacks will allow individuals to enter *personal subscriptions.*

MEDICAL ELECTROENCEPHALOGRAPHY

by FREDERIC A. GIBBS and ERNA L. GIBBS
University of Illinois, College of Medicine

From the Authors' Preface:
"Most laymen and many physicians on receiving a report from the electroencephalographic laboratory are dismayed by their inability to interpret it. They ask, 'What does mean?' 'Is it normal?' or 'How abnormal is it?' This book explains what physicians and other interested persons will want to know about the meaning of encephalographic findings.

"This concise book is designed for physicians who refer patients for encephalographic study, and for psychologists, social workers, medical students, and nurses who need to know the clinical implications of electroencephalography. It will also serve as a valuable reference work for EEG departments and as a supplementary text for courses in neurology. Although the book is based on statistical data, the presentation is an nontechnical as possible. Plain English and simple mathematical expressions are used throughout."
1967, viii, 79 pp., illus., 8-1/4" X 11", hardbound ISBN 0-201-02365-2

Addison-Wesley Publishing Company, Inc.
Advanced Book Program
Reading, Massachusetts 01867

Second Edition
Completely revised, reset, and enlarged

BIOLOGICAL TRANSPORT
by Halvor N. Christensen

An updated presentation of central aspects of molecular transport across biological membranes is now available in this completely revised edition. The parallel subject of group translocation is discussed. Here is a critical guide to the original literature for use by the advanced student of bioscience (biophysics, biochemistry, physiology). Investigators in these subjects and in microbiology, physiology, pharmacology, genetics, and other researchers working on membrane and transport will find the volume of special interest.

Halvor N. Christensen holds degrees of B.S. in Chemistry from Kearney State College, M.S. and Ph.D. in Biochemistry from Purdue and Harvard Universities, respectively, and is currently professor in the Department of Biological Chemistry at the University of Michigan. His previous assignments include working as a biochemist at Lederle Laboratories; Assistant Professor at Harvard University; Director, Chemistry Laboratory, Mary Imogene Bassett Hospital, New York; Director, Chemical Research, Children's Medical Center, Boston; Assistant Professor, Harvard Medical School; Professor and Chairman, Biological Chemistry and Nutrition, Tufts Medical School; Chairman, Biological Chemistry, University of Michigan.

The author has been a full-time student of the field of biological transport for a quarter of a century, and with this volume, seeks to compensate for fragmentation of the subject among disciplines, with particular emphasis on molecular approaches.

1975, xx, 514 pp., illus., hardbound ISBN 0-805-32251-5

W. A. BENJAMIN, INC.
Advanced Book Program
Reading, Massachusetts